# MATH FOR KIDS

## Ages: 4-5-6 years old

More than **150** exercises

**Counting, order, addition, subtraction, pattern, geometry, critical thinking, problem solving, reasoning ability and the ability to communicate mathematically are essential skills in this workbook.**

## Note for parents

Parents are an important partner in mathematics education.

We must built positive attitudes about math. We must encourage child and working with him or her. We must find ways to engage child in thinking and talking about mathematics, for his or her future success.

## Author :

Sofian Makhlouf is teacher of mathematics with more than twenty years of experience.

# Contents

# Chapter 1

# Coloring and geometry

Exercise 1 : Trace the lines with pen or a pencel.

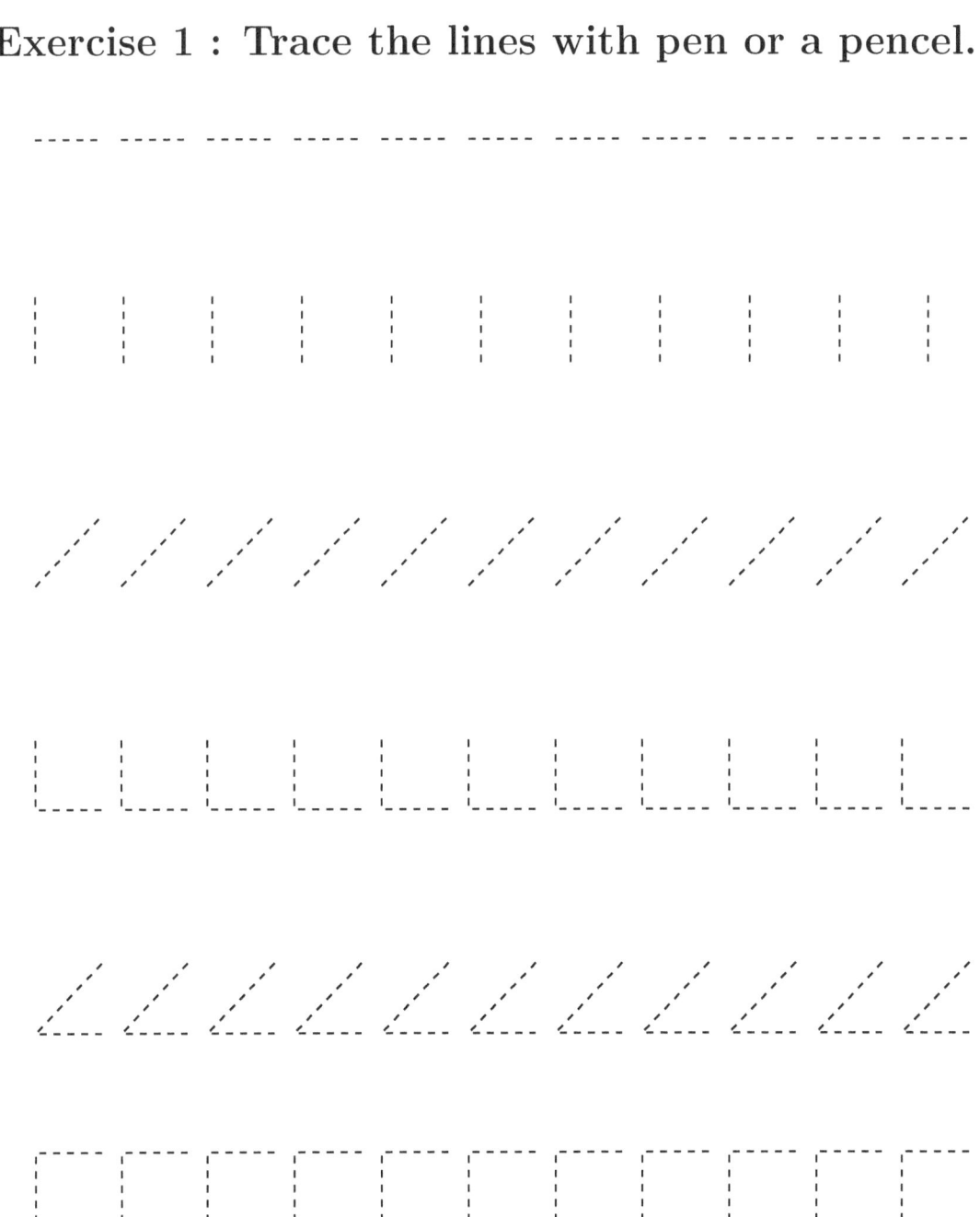

**Exercise 2 :** **Color the triangles whith red.**

**Exercise 3 :** **Color the circles with yellow.**

**Exercise 4 :** **Color the squares whith blue.**

**Exercise 5 :** **Color the rectangles whith green.**

**Exercise 6 :** **Color the stars whith orange.**

# Exercise 7 : Match and color.

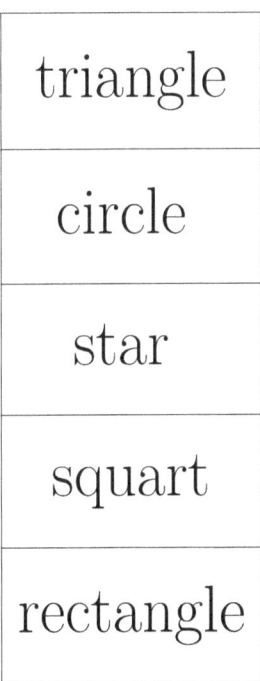

triangle

circle

star

squart

rectangle

**Exercise 8 : Help Jhon to take the balloon and color the balloon.**

Jhon

**Exercise 9 :** Help the cat to take the fish and color the fish.

**Exercise 10 : Color the car.**

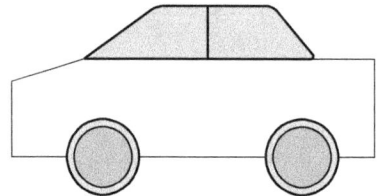

**Exercise 11 : Color the trees.**

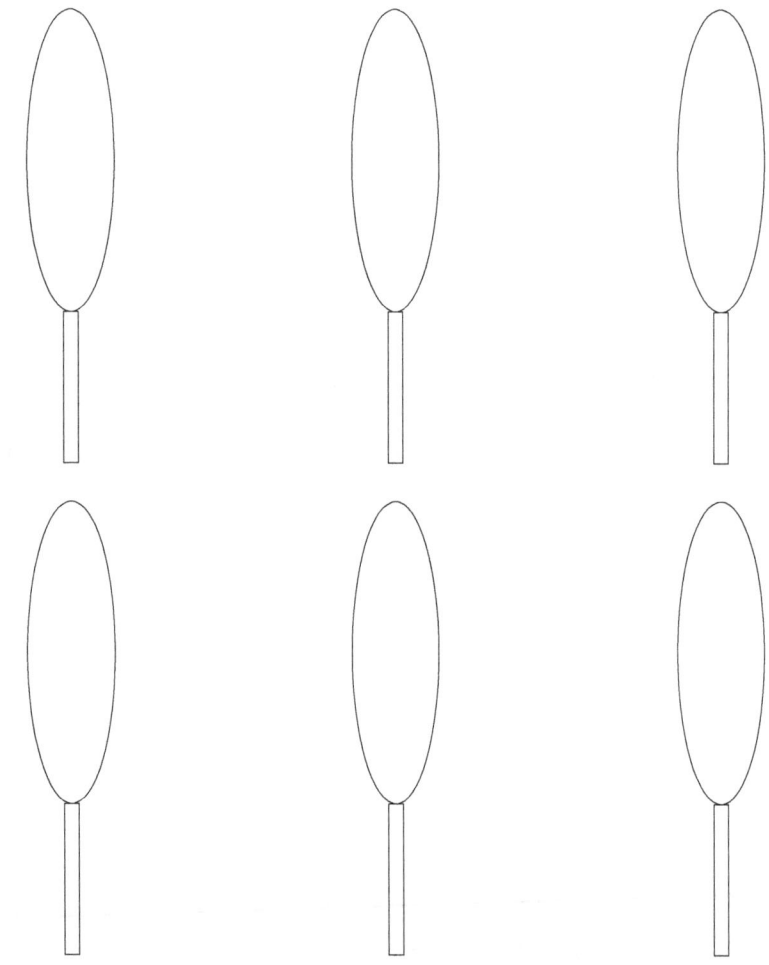

# Chapter 2

# Thinking

## 2.1 Similar and pattern.

Exercise 1 : Find similar shapes below and tick.

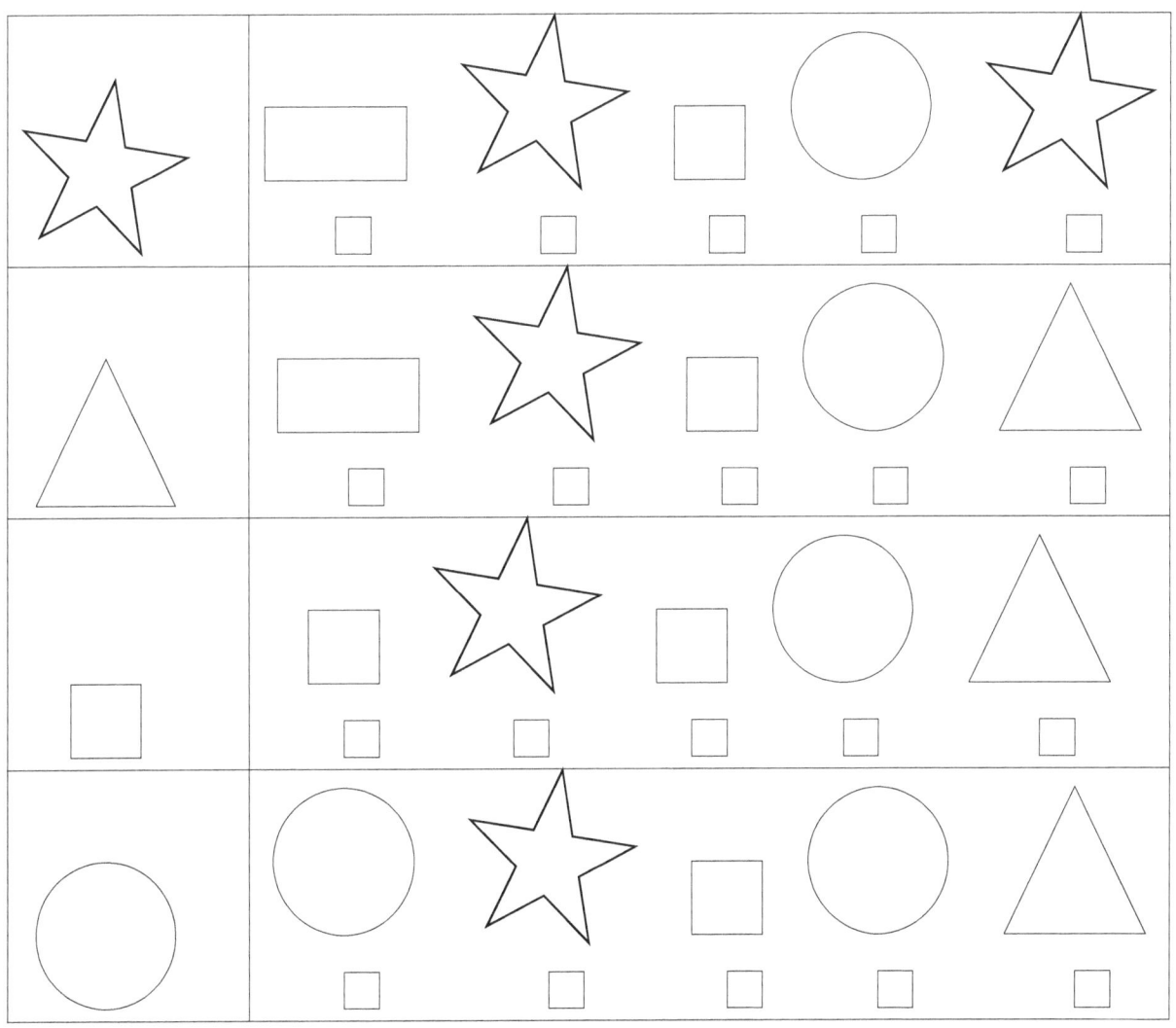

**Exercise 2 : Sort out and tick the pattern on the right that is similar to that on the left.**

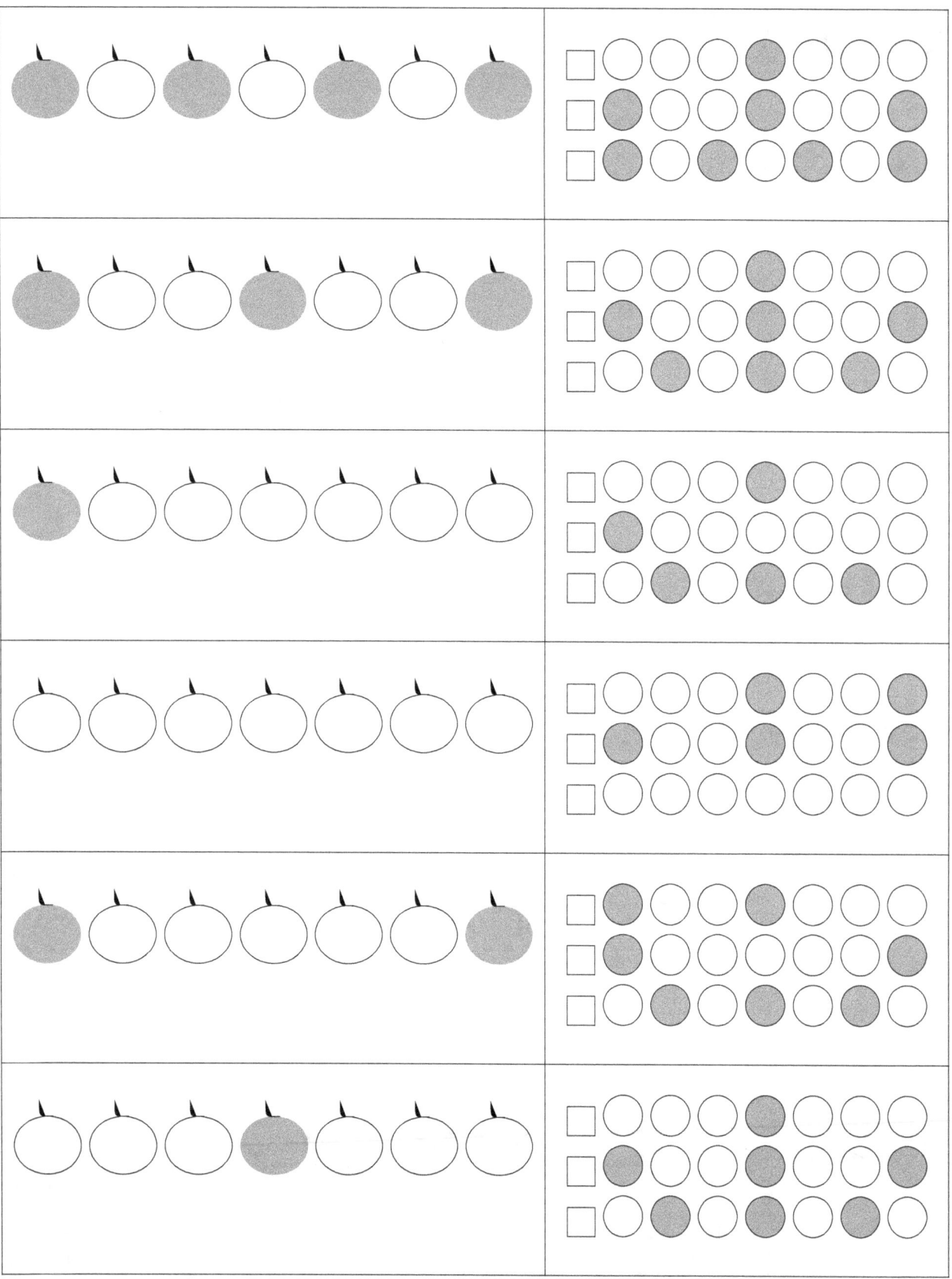

# Exercise 3 : Sort out and tick the pattern on the right that comes next.

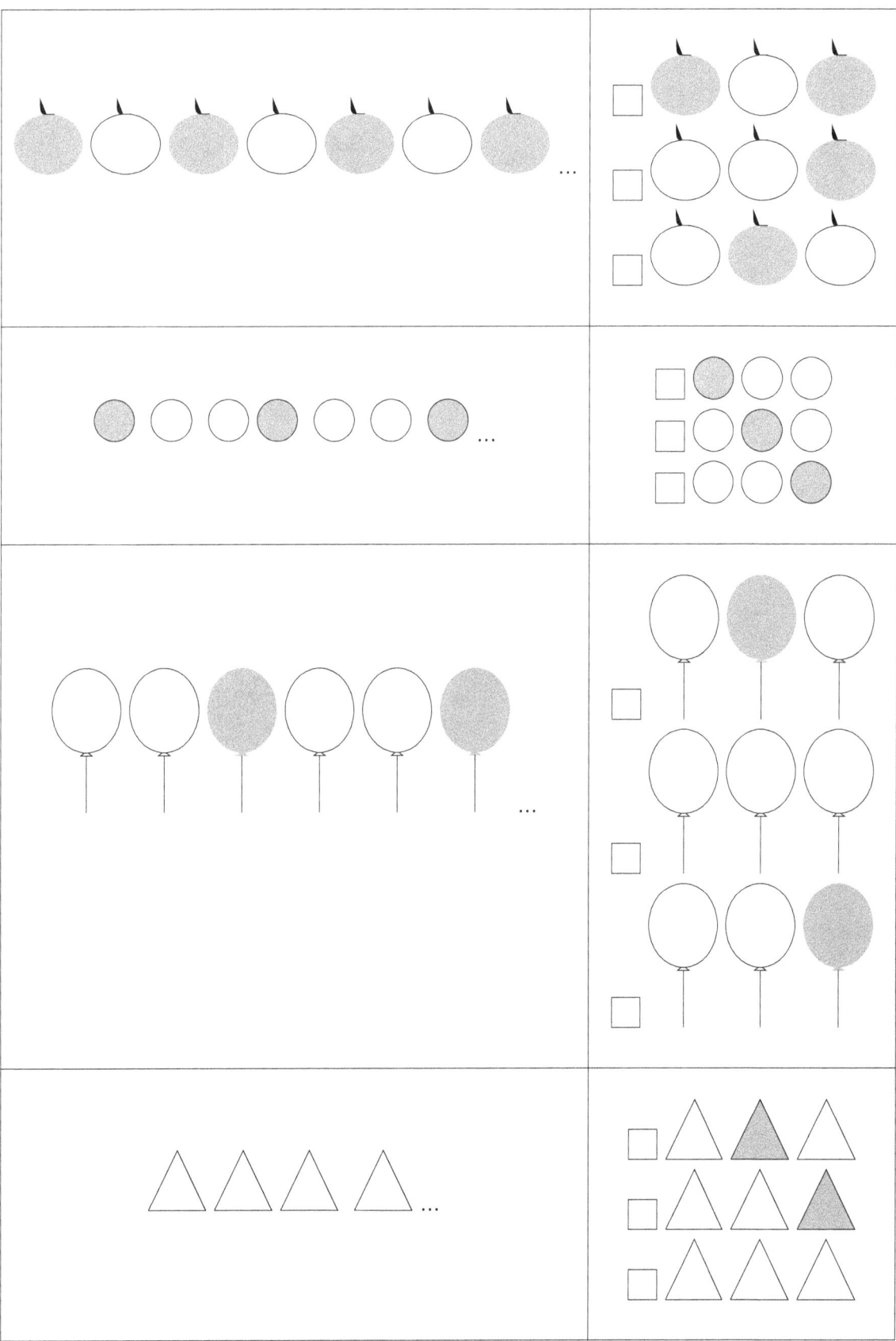

**Exercise 4 : Sort out and tick shapes that are different below.**

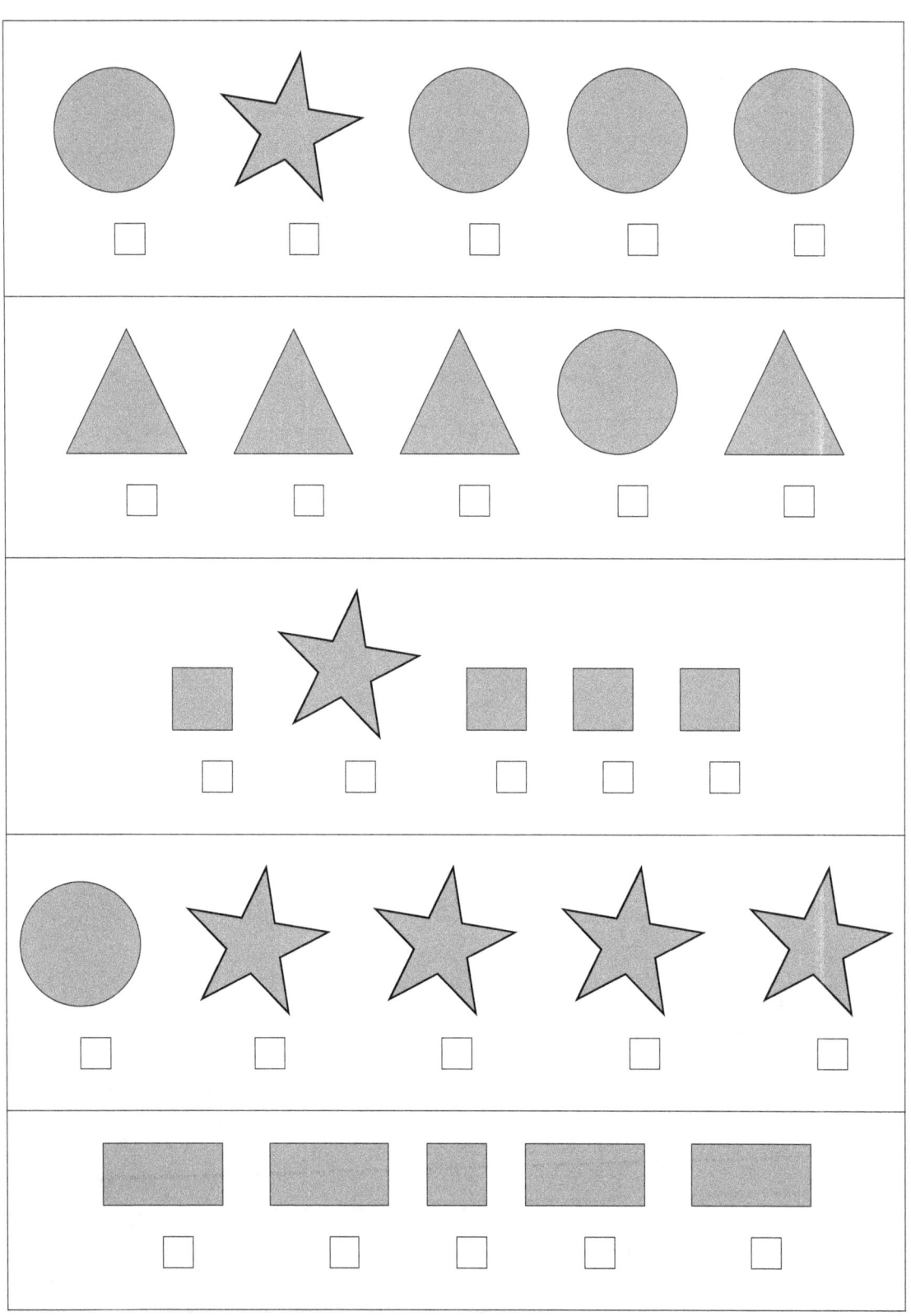

## 2.2   Big and small

The white circle is **smaller than**  the gray circle .

The gray circle is **bigger than** the white circle.

**Exercise 5 : Fill in the blanks with small, smaller, big and bigger.**

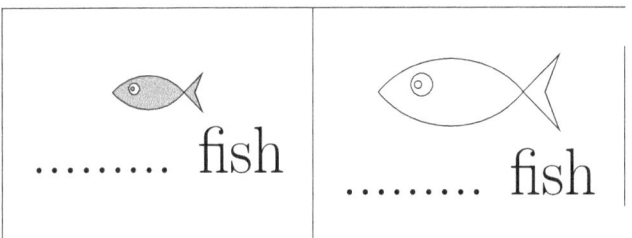

The white fish is ............... than the gray fish.

The gray fish is .............. than the white fish.

## 2.3 Tall and short

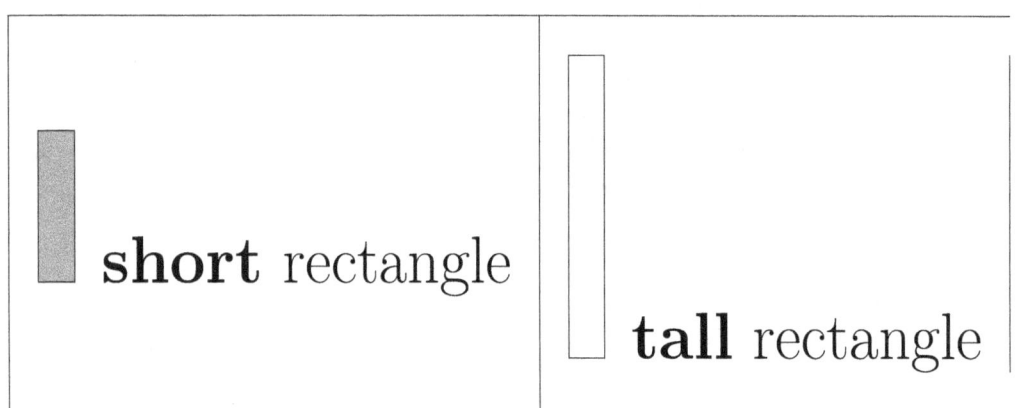

The gray rectangle is **shorter than** the white rectangle.

The white rectangle is **taller than** the gray rectangle.

Jhon is **shorter than** Sam.

Sam is **taller than** Jhon.

**Exercise 6 : Fill in the blanks with short, shorter, tall and taller.**

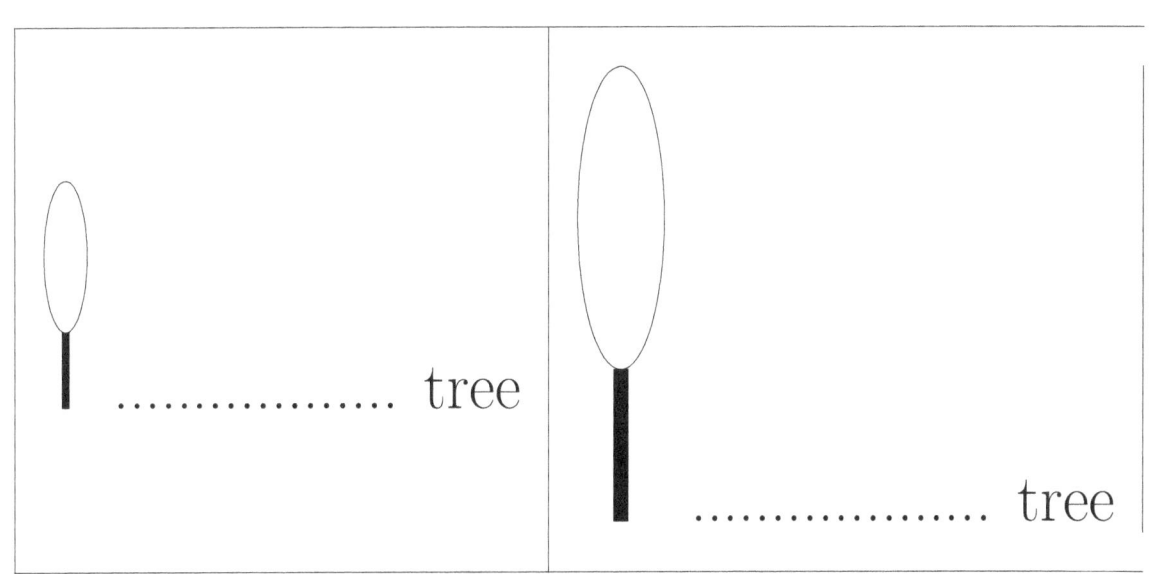

...................... tree

...................... tree

The tree on the left is .............. than the tree on the right.

The tree on the right is .............. than the tree on the left.

...................... ladder

...................... ladder

The ladder on the left is .............. than the ladder on the right.

The ladder on the right is .............. than the ladder on the left.

# Exercise 7 : Help Jhon to take the shortest way to balloon.

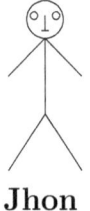

Jhon

# Exercice 8 : Help Jhon and Victor to take the longest way to home.

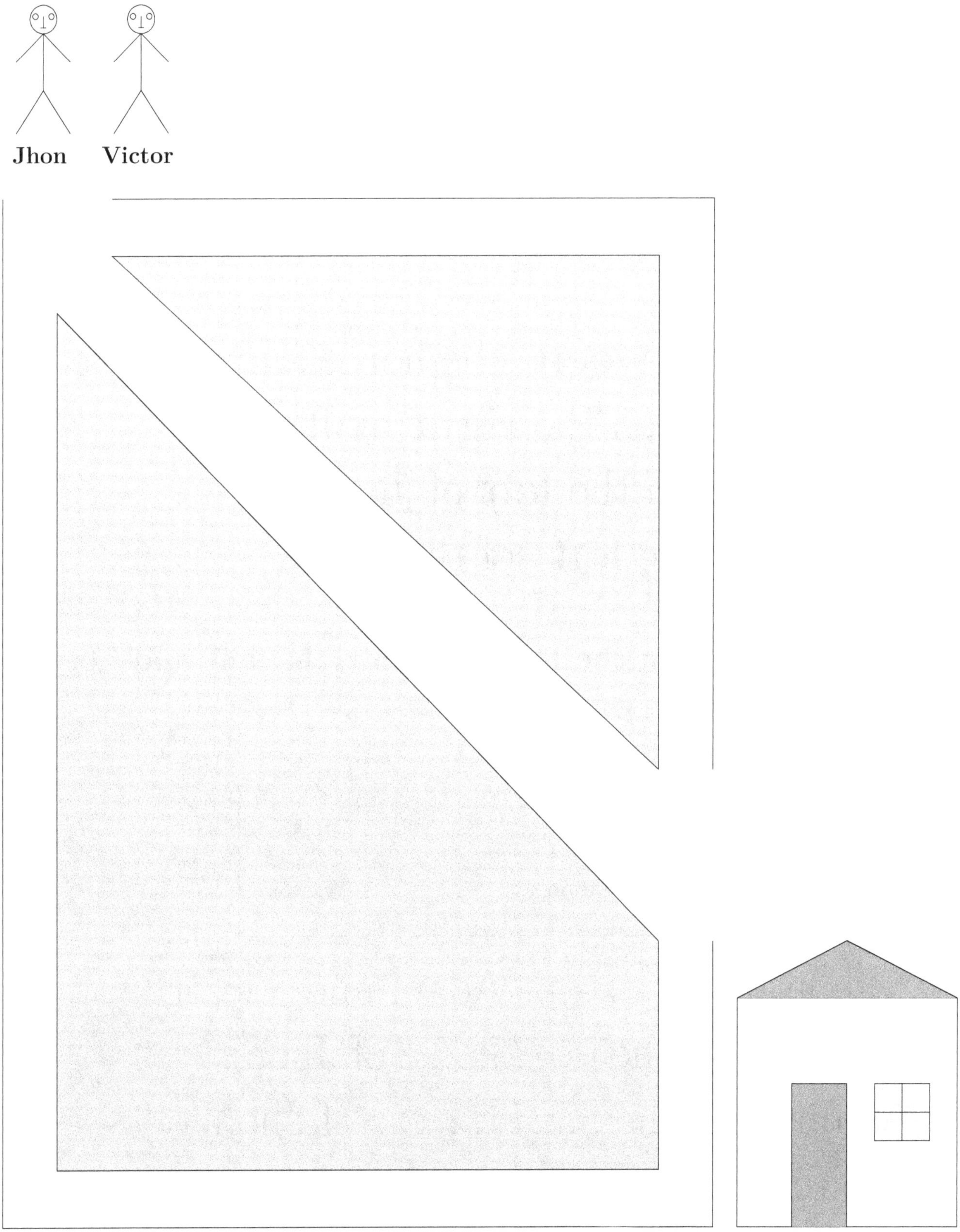

Jhon    Victor

## 2.4   Right, left and between

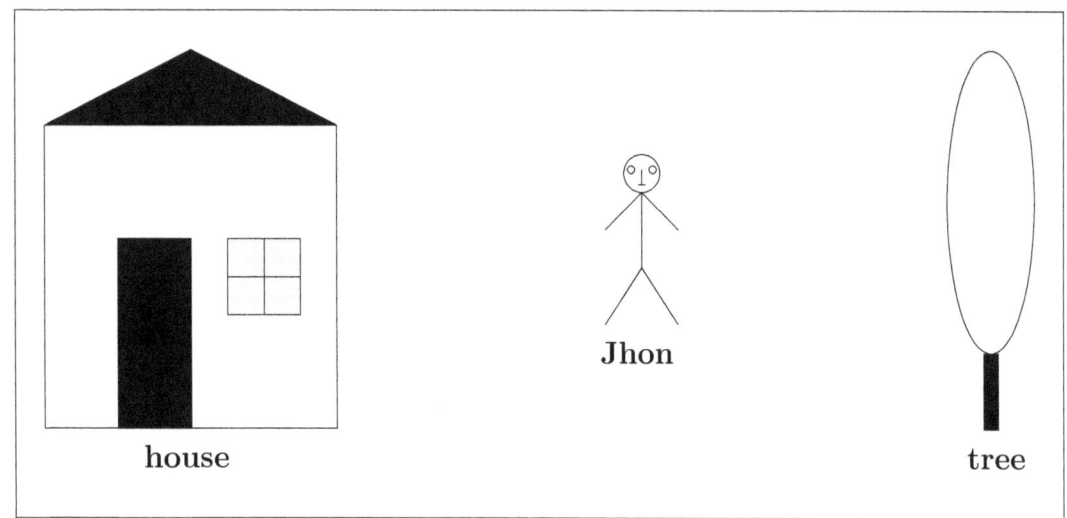

- Jhon is **between**  the house and the tree.
- The house is on the **right** of Jhon.
- The tree is on the **left** of Jhon.
- Jhon is on the **left**  of the house.

**Exercise 9: Fill in the blanks with right, left and between.**

- Jhon is ................ Adam and Victor.
- Victor is on the ............... of Jhon.
- Adam is on the ................ of Jhon.
- Jhon is on the ................ of Adam.

## Exercise 10 : Fill in the blanks with right, left and between

- Jhon is ................ Adam and Victor.
- Victor is on the ............... of Jhon.
- Adam is on the .................. of Jhon.
- Jhon is on the ................. of Adam.

## Exercise 11: fill in the blanks with right, left and between

- Jhon is ................ Adam and Victor.
- Victor is on the ............... of Jhon.
- Adam is on the ................... of Jhon.
- Jhon is on the ................. of Adam.
- Victor is on the ................. of Adam.
- Jhon is on the ................. of Victor.

**Exercise 12 : Fill in the blanks with right or left.**

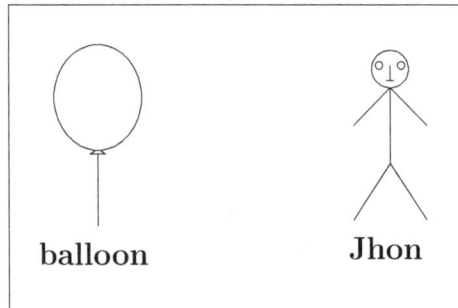

• Balloon is on the ............... of Jhon.

The balloon is on my left.

**Exercise 13 : Fill in the blanks with right or left.**

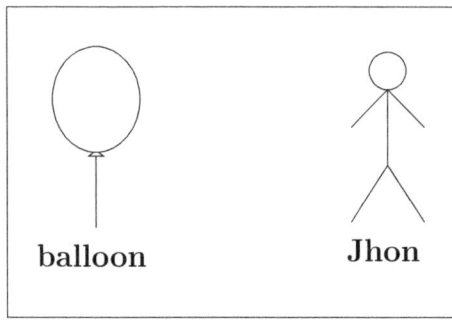

• Balloon is on the ............... of Jhon.

**Exercise 14 : Fill in the blanks with right or left.**

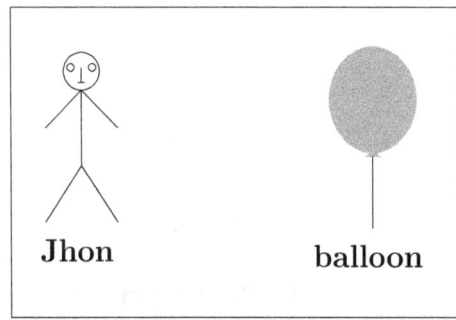

• Balloon is on the ............... of Jhon.

**Exercise 15 : fill in the blanks with right or left.**

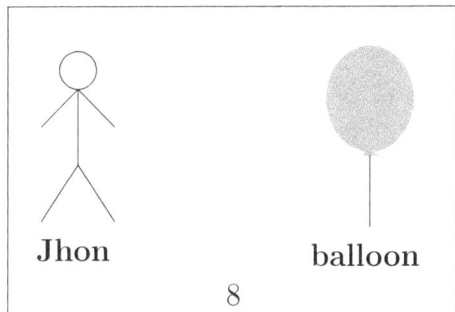

. Balloon is on the ............... of Jhon.

The balloon is on my right.

**Exercise 16 : fill in the blanks with right, between or left.**

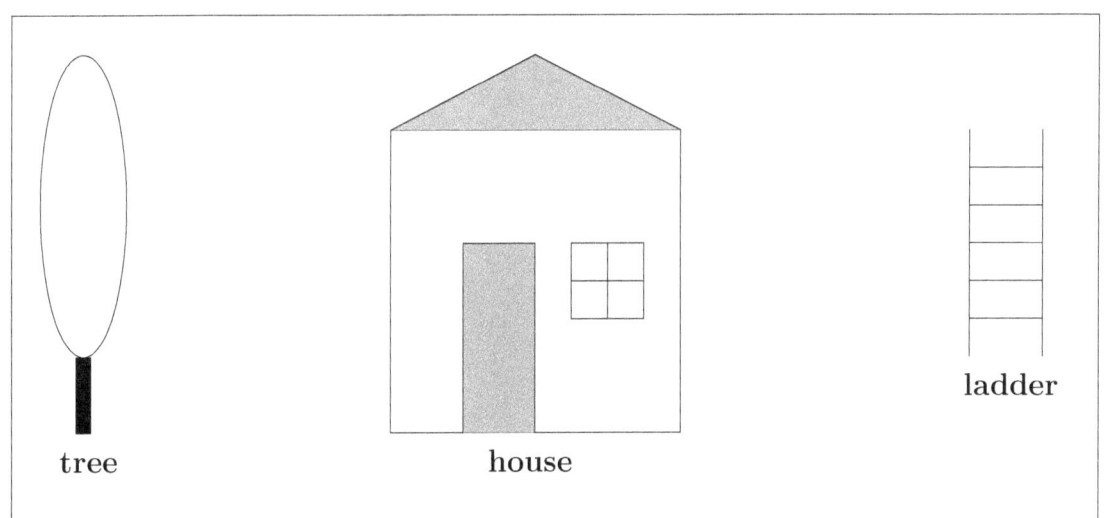

. The house is ............. the hadder and the tree.

. The hadder is on the ............ of house.

. The tree is on the ........... of the house.

**Exercise 17 : fill in the blanks with right, between or left.**

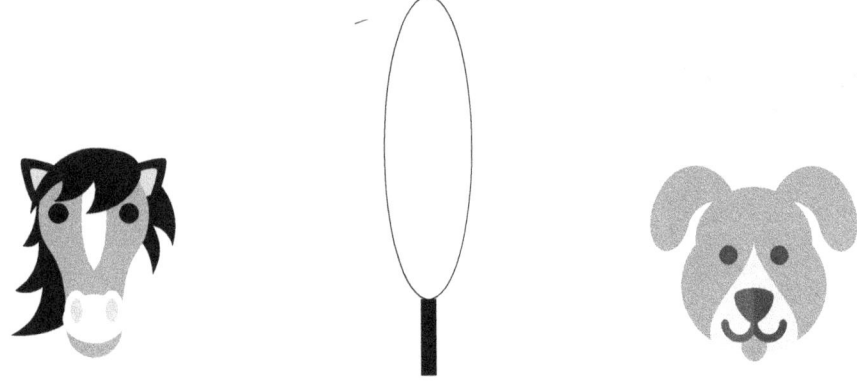

• The tree is ............ the horse and the dog.

• The tree is on the ........... of the horse.

• The horse is on the ........... of the dog.

**Exercise 18 : fill in the blanks with right, between or left.**

• The horse is ............ the cat and the dog.

• The cat is on the ........... of the horse.

• The horse is on the ........... of the dog.

• The dog is on the ........... of the cat .

# Chapter 3

# Numbers counted up to five

## 3.1  Numbers counted up to three

| | | |
|---|---|---|
| | 1 | one dog |
| | 2 | two dogs |
| | 3 | three dogs |

| | | |
|---|---|---|
| | 1 | one shark |
| | 2 | two sharks |
| | 3 | three sharks |

| | | |
|---|---|---|
| | 1 | one ladybird |
| | 2 | two ladybirds |
| | 3 | three ladybirds |

**Exercise 1:** Trace the number using a pencil or pen.

0

1

2

3

**Exercise 2:** Complete.

| | | 0 | zero |
|---|---|---|---|
| • | | | one |
| • • | | 2 | |
| • • • | | | three |
| ♡ ♡ ♡ | | 3 | |
| ♡ ♡ | | | two |
| ♡ | | 1 | |

# Exercise 3: How many things are there?

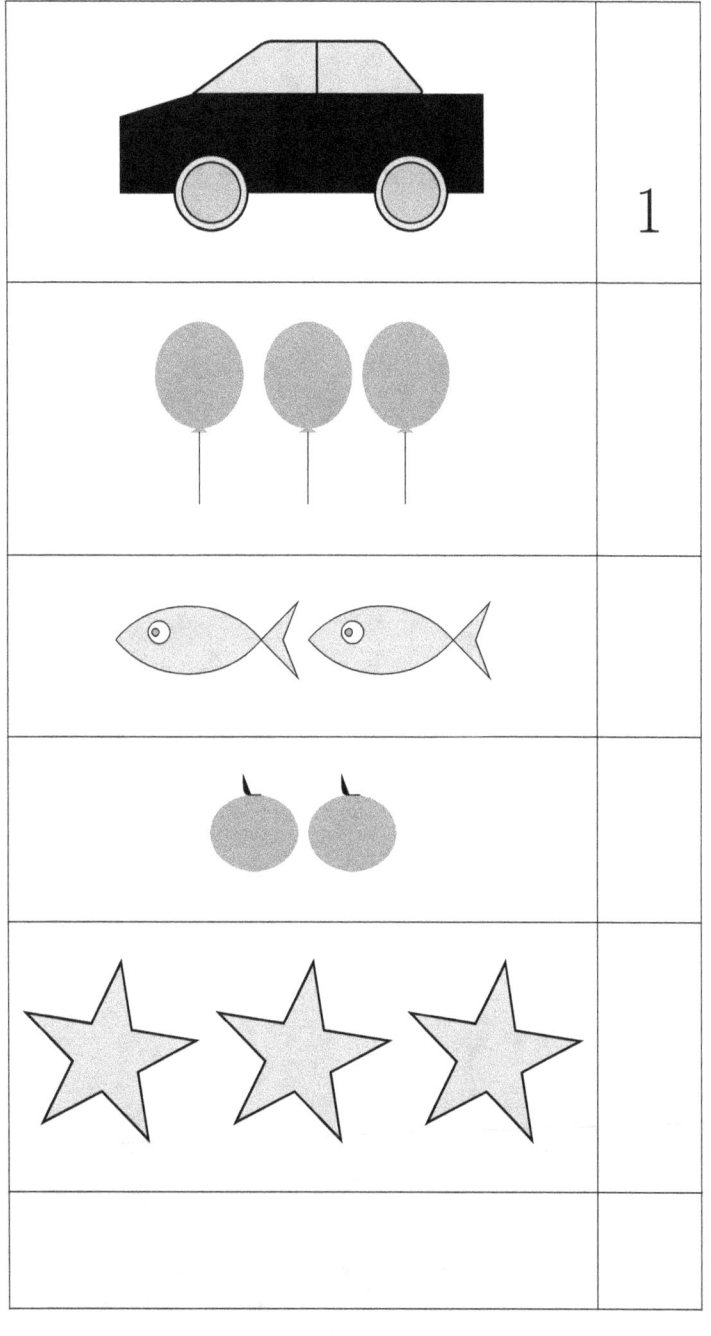

## Exercise 4 : Match.

| | |
|---|---|
| 0 | •• two dots |
| 1 | ••• three dots |
| 2 | zero dot |
| 3 | • one dot |

## Exercise 5

| | |
|---|---|
| Draw 1 circle | |
| Draw 2 circles | |
| Draw 3 circles | |

## Exercise 6

| | |
|---|---|
| Draw 1 squart | |
| Draw 2 squarts | |
| Draw 3 squarts | |

# Exercise 7

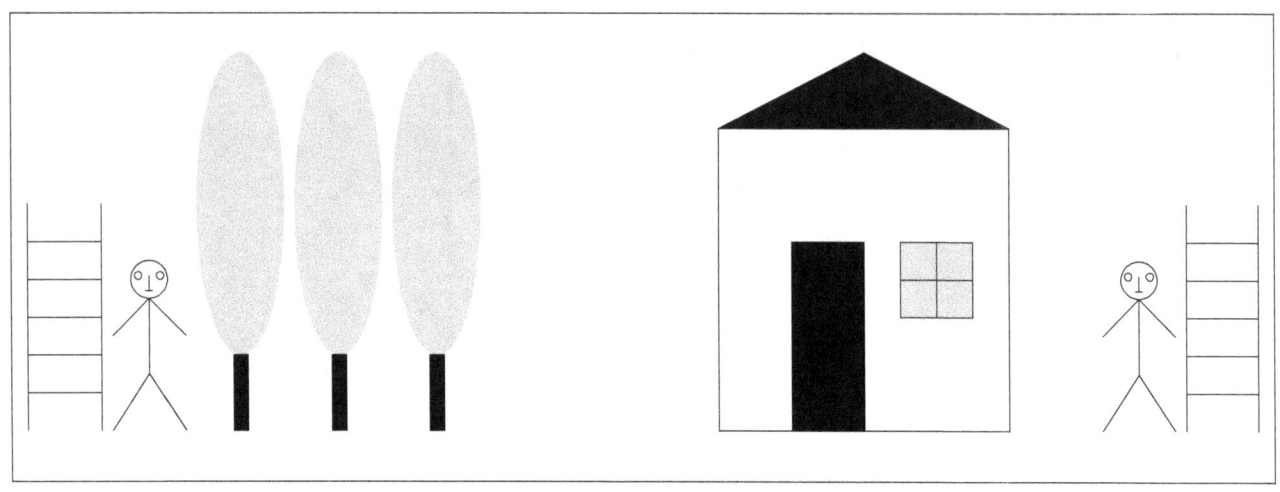

| | |
|---|---|
| How many houses are there? | |
| How many trees are there? | |
| How many hadders are there? | |
| How many peoples are there? | |

**Exercise 8 : How many cats are there ?**

        There are ... cats.

## 3.2 Numbers counted up to five

| | | |
|---|---|---|
| | 4 | four cats |
| | 5 | five cats |

| | | |
|---|---|---|
| | 4 | four apples |
| | 5 | five apples |

## Exercise 9 : Trace the number using a pencil or pen.

4

5

## Exercise 10 : Choose the name to number.

| | |
|---|---|
| 0 | two |
| 1 | four |
| 2 | three |
| 3 | five |
| 4 | zero |
| 5 | one |

## Exercise 11 : Complete the tables.

| | | |
|---|---|---|
| | 0 | zero |
| • | | one |
| •• | 2 | |
| ••• | | three |
| •••• | | four |
| ••••• | | five |

| | | |
|---|---|---|
| ♡ ♡ ♡ ♡ ♡ | 5 | |
| ♡ ♡ ♡ ♡ | 4 | |
| ♡ ♡ ♡ | 3 | |
| ♡ ♡ | | two |
| ♡ | 1 | |
| | | |

# Exercise 12 : How many things are there?

| | |
|---|---|
| ‖‖‖‖ | 4 |
| 🎈🎈🎈🎈🎈 | |
| └ └ └ | |
| ☆☆☆☆☆ | |
| ▲▲▲ | |
| ▮▮▮▮ | |
| ▭▭ | |

# Exercise 13

| | |
|---|---|
| How many eyes are there ? | 2 |
| How many head are there ? | |
| How many arms are there ? | |
| How many legs are there ? | |
| How many fingers are there ? | |

# Exercise 14

.How many houses are there ?

.How many cars are there ?

.How many peoples on the right of house are there ?

.How many peoples are there ?

.How many trees on the right of house are there ?

.How many trees on the left of house are there ?

.How many trees are there ?

**Exercise 15**

. How many white balloons are there ?

. How many balloons on the right of Jhon are there ?

. How many balloons on the left of Jhon are there ?

. How many gray balloons are there ?

. How many big balloons are there ?

. How many small balloons are there ?

. How many green balloons are there ?

. How many balloons are there ?

## 3.3  Number after before and between.

**1 2 3 4 5**

The number comes between 1 and 3 is 2.

The number comes after 2 is 3.

The number comes before 3 is 2.

**Exercise 16**

| | |
|---|---|
| Which number comes between 3 and 5 ? | |
| Which number comes after 2 ? | |
| Which number comes after 3 ? | |
| Which number comes before 4 ? | |
| Which number comes before 5 ? | |

**Exercise 17 : Write the number comes after.**

| 1 | 2 | |
|---|---|---|

| 2 | 3 | |
|---|---|---|

| 3 | 4 | |
|---|---|---|

**Exercise 18 :** Write the number comes before.

|   | 2 | 3 |
|---|---|---|

|   | 3 | 4 |
|---|---|---|

|   | 4 | 5 |
|---|---|---|

**Exercise 19 :** Write the number comes between.

| 1 |   | 3 |
|---|---|---|

| 3 |   | 5 |
|---|---|---|

| 2 |   | 4 |
|---|---|---|

**Exercise 20 :** Write the missing number.

| 1 | 2 | 3 | 4 |   |
|---|---|---|---|---|

| 1 | 2 | 3 |   | 5 |
|---|---|---|---|---|

| 1 | 2 |   | 4 | 5 |
|---|---|---|---|---|

| 1 |   | 3 | 4 | 5 |
|---|---|---|---|---|

|   | 2 | 3 | 4 |   |
|---|---|---|---|---|

| 1 |   | 3 |   | 5 |
|---|---|---|---|---|

|   | 2 |   | 4 |   |
|---|---|---|---|---|

|   |   | 3 |   | 5 |
|---|---|---|---|---|

| 1 |   |   | 4 |   |
|---|---|---|---|---|

| 1 |   |   |   | 5 |
|---|---|---|---|---|

| 1 |   |   | 4 |   |
|---|---|---|---|---|

|   |   |   |   | 5 |
|---|---|---|---|---|

## 3.4 Ordinal : first, seconde, third, fourth and fifth.

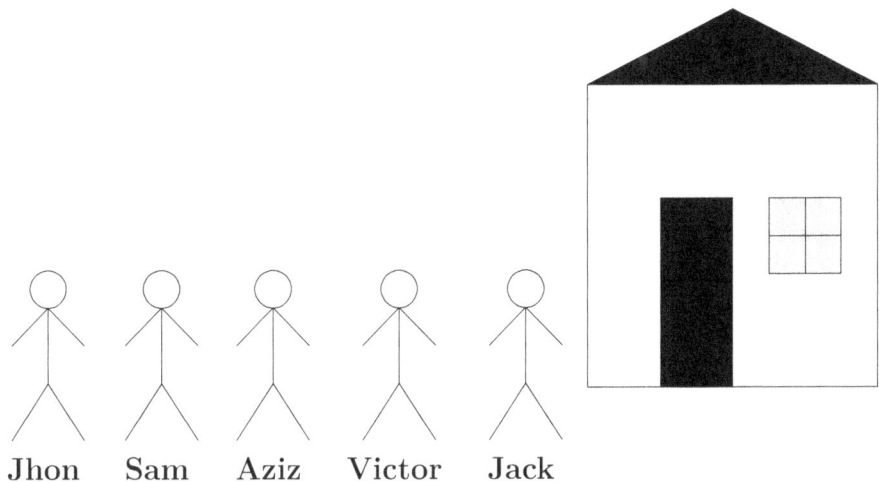

- Jack arrive at home first.
- Victor arrive at home second.
- Aziz arrive at home third.
- Sam arrive at home fourth.
- Jhon arrive at home fifth.

| Cardinal | |
|---|---|
| 1 | one |
| 2 | two |
| 3 | three |
| 4 | four |
| 5 | five |

| Ordinal | |
|---|---|
| 1st | first |
| 2nd | second |
| 3rd | third |
| 4th | fourth |
| 5th | fifth |

**Exercise 21 :** Circle the third duck.

**Exercise 22 :** Circle the fifth shark.

**Exercise 23 :** Circle the second bee.

# Chapter 4

# Numbers counted up to ten

## 4.1 Five and six

| | | |
|---|---|---|
| | 5 | five apples |
| | 6 | six apples |

| | | |
|---|---|---|
| | 5 | five balloons |
| | 6 | six balloons |

**Exercise 1 : Trace the numbers using a pencil or pen.**

5     5     5     5     5     5

6     6     6     6     6     6

**Exercise 2 : How many trees are there ?.**

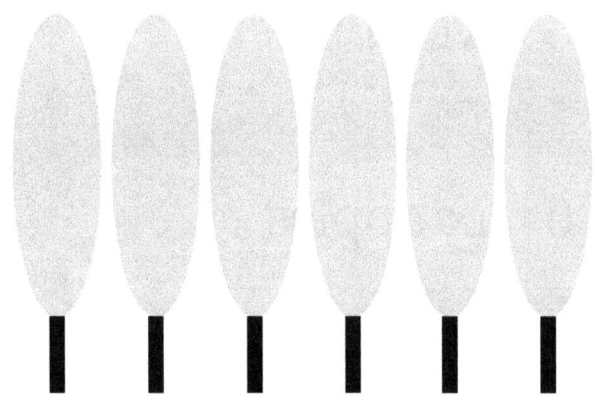

Ther are ... trees.

**Exercise 3 : How many circles are there ?.**

Ther are ... circles.

## 4.2 Seven and eight

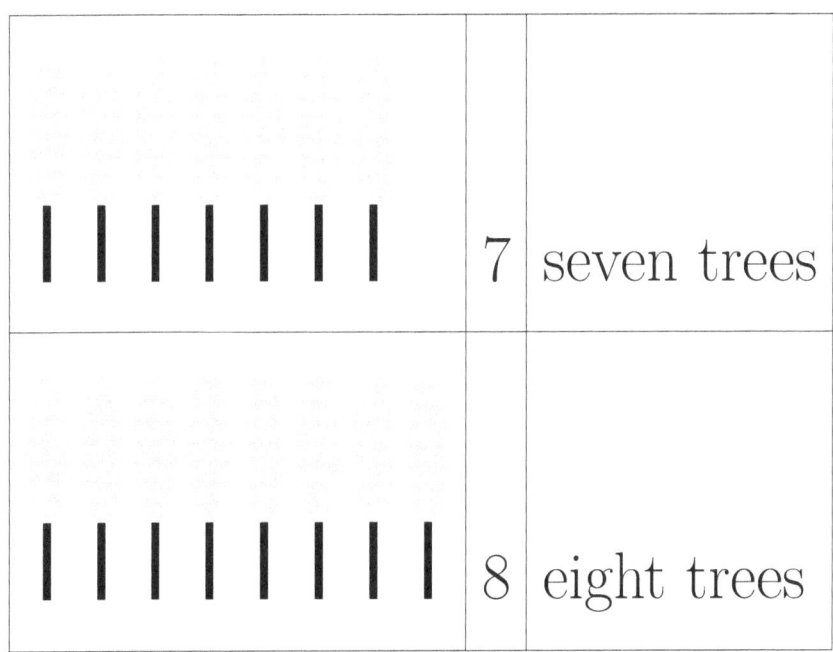

| | 7 | seven trees |
| 8 | eight trees |

| 7 | seven circles |
| 8 | eight circles |

**Exercise 4 : Trace the number using a pencil or pen.**

7

8

## Exercise 5 : How many apples are there ?

| | |
|---|---|
| | |
| 🍎 | |
| 🍎 🍎 | |
| 🍎 🍎 🍎 | |
| 🍎 🍎 🍎 🍎 | |
| 🍎 🍎 🍎 🍎 🍎 | |
| 🍎 🍎 🍎 🍎 🍎 🍎 | |
| 🍎 🍎 🍎 🍎 🍎 🍎 🍎 | |
| 🍎 🍎 🍎 🍎 🍎 🍎 🍎 🍎 | |

## 4.3 Nine and ten

| | 9 | nine |
|---|---|---|
| | 10 | ten |

| | 9 | nine |
|---|---|---|
| | 10 | ten |

**Exercise 6 : Trace the number using a pencil or pen.**

9

10

# Exercise 7 : Complete the tables.

| | | |
|---|---|---|
| • • | 2 | |
| • • • | | three |
| • • • • | | four |
| • • • • • | | five |
| • • • • • • | | six |
| • • • • • • • | 7 | |
| • • • • • • • • | | eight |
| • • • • • • • • • | | nine |
| • • • • • • • • • • | | ten |

| | | |
|---|---|---|
| ♡ ♡ ♡ ♡ ♡ ♡ ♡ ♡ ♡ ♡ | 10 | |
| ♡ ♡ ♡ ♡ ♡ ♡ ♡ ♡ ♡ | 9 | |
| ♡ ♡ ♡ ♡ ♡ ♡ ♡ ♡ | 8 | |
| ♡ ♡ ♡ ♡ ♡ ♡ ♡ | | seven |
| ♡ ♡ ♡ ♡ ♡ ♡ | 6 | |
| ♡ ♡ ♡ ♡ ♡ | 5 | |
| ♡ ♡ ♡ ♡ | 4 | |
| ♡ ♡ ♡ | 3 | |
| ♡ ♡ | | two |

**Exercise 8 : Match.**

| | | |
|---|---|---|
| • • | 0 | three |
| | 1 | zero |
| • | 2 | one |
| • • • • | 3 | two |
| • • • • • • | 4 | six |
| • • • • • • • | 5 | nine |
| • • • | 6 | ten |
| • • • • • | 7 | four |
| • • • • • • • • | 8 | eight |
| • • • • • • • | 9 | seven |
| • • • • • • • • | 10 | five |

I have ten fingers in my two hands.

# Exercise 9 : How many circles are there ?

| | |
|---|---|
| ● ● ● ● ● ● | |
| ● ● ● ● ● ● ● ● | |
| ● ● ● ● ● | 5 |
| ● ● ● ● ● ● ● | |
| ● ● ● | |
| ● ● ● ● ● ● ● ● ● ● | |
| ● ● ● ● ● | |
| ● ● | |
| ● ● ● ● ● ● ● ● | |
| ● ● ● ● ● ● ● ● ● ● | |

# Exercise 10

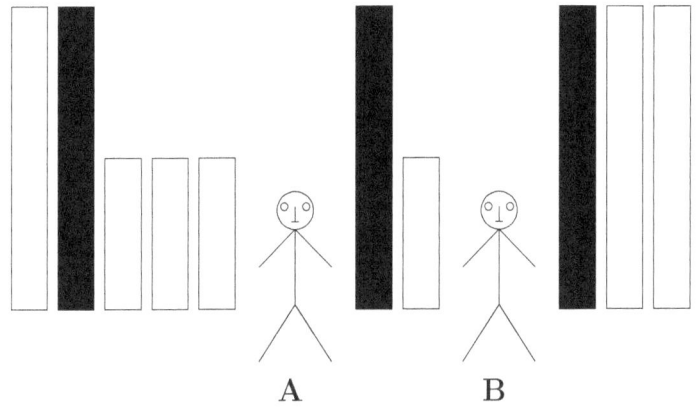

| | |
|---|---|
| How many black rectangles are there ? | 3 |
| How many white rectangles are there ? | |
| How many rectangles are there ? | |
| How many short rectangles are there ? | |
| How many tall rectangles are there ? | |
| How many green rectangles are there ? | |
| How many rectangles between A and B ? | |
| How many rectangles on the right of B ? | |
| How many rectangles on the left of A ? | 5 |

**Exercise 11**

| | |
|---|---|
| How many yellow balloons are there ? | |
| How many gray balloons are there ? | |
| How many big gray balloons are there ? | |
| How many white balloons are there ? | |
| How many big balloons are there ? | |
| How many small balloons are there ? | |
| How many balloons are there ? | |
| How many balloons on the right of Jhon ? | |
| How many balloons on the left of Jhon ? | |

**Exercise 12 : How many letters are ther in each world ?**

| | |
|---|---|
| in | 2 |
| are | |
| car | |
| hand | |
| star | |
| letter | |
| number | |
| fish | |
| peoples | |
| rectangle | |
| balloon | |
| yellow | |
| color | |

| | |
|---|---|
| the | |
| things | |
| rectangles | |
| house | |
| robot | |
| between | 7 |
| left | |
| right | |
| tall | |
| short | |
| big | |
| small | |
| green | |

## 4.4   Number after before and between.

**1 2 3 4 5 6 7 8 9 10**

The number comes between 5 and 7 is 6.

The number comes after 6 is 7.

The number comes before 7 is 6.

### Exercise 13

| | |
|---|---|
| Which number comes between 4 and 6 ? | |
| Which number comes between 7 and 9 ? | |
| Which number comes after 5? | |
| Which number comes after 7 ? | |
| Which number comes after 8 ? | |
| Which number comes before 5 ? | |
| Which number comes before 10 ? | |

**Exercise 14 : Write the number comes after.**

| 4 | 5 | |

| 8 | 9 | |

| 6 | 7 | |

**Exercise 15 : Write the number comes before.**

| | 6 | 7 |
|---|---|---|

| | 7 | 8 |
|---|---|---|

| | 8 | 9 |
|---|---|---|

| | 5 | 6 |
|---|---|---|

| | 9 | 10 |
|---|---|---|

| | 2 | 3 |
|---|---|---|

**Exercise 16 : Write the number comes between.**

| 4 | | 6 |
|---|---|---|

| 5 | | 7 |
|---|---|---|

| 6 | | 8 |
|---|---|---|

| 7 | | 9 |
|---|---|---|

| 2 | | 4 |
|---|---|---|

| 8 | | 10 |
|---|---|---|

**Exercise 17 : Write the missing number.**

| 1 | | 3 | | 5 | | 7 | | 9 | |
|---|---|---|---|---|---|---|---|---|---|

| | 2 | | 4 | | 6 | | 8 | | 10 |
|---|---|---|---|---|---|---|---|---|---|

## 4.5   Ordinal numbers

| Cardinal | | Ordinal | |
|---|---|---|---|
| 1 | one | 1st | first |
| 2 | two | 2nd | second |
| 3 | three | 3rd | third |
| 4 | four | 4th | fourth |
| 5 | five | 5th | fifth |
| 6 | six | 6th | sixth |
| 7 | seven | 7th | seventh |
| 8 | eight | 8th | eighth |
| 9 | nine | 9th | ninth |
| 10 | ten | 10th | tenth |

The first letter of the word numbers is n.

The second letter of the word numbers is u.

The third letter of the word numbers is m.

The fourth letter of the word numbers is b.

The fifth letter of the word numbers is e.

**Exercise 18 : What is the first letter of each word below?**

| | | |
|---|---|---|
| | dog | |
| | apple | a |
| | circle | |
| | hand | |
| | star | |
| | triangle | |
| | fish | |

**Exercise 19 : What is the first letter of each word below?**

| | | | | |
|---|---|---|---|---|
| in | i | many | |
| are | | things | |
| cat | | before | |
| ordinal | | what | |
| square | | robot | |
| how | | triangles | t |
| number | | dog | |
| first | | left | |
| pen | | right | |
| rectangle | | vertical | |
| book | | play | |

**Exercise 20 : Fill in the blanks.**

☐ ☐ two rectangles

How many letters in word rectangles are there ?

The first letter of the word rectangles is ....

The second letter of the word rectangles is ....

The third letter of the word rectangles is ....

The fourth letter of the word rectangles is ....

The sixth letter of the word rectangles is ....

The seventh letter of the word rectangles is ....

The ninth letter of the word rectangles is ....

The tenth letter of the word rectangles is ....

**Exercise 21 : Fill in the blanks.**

There are seven days in a week : monday, tuesday, wednesday, thursday, friday, saturday and sunday.

The first day in a week is monday

The seconde day in a week is tuesday

The third day in a week is ...................

The fourth day in a week is ...................

The fifth day in a week is ...................

The sixthth day in a week is ...................

The seventh day in a week is ...................

# Chapter 5

# Comparison of numbers

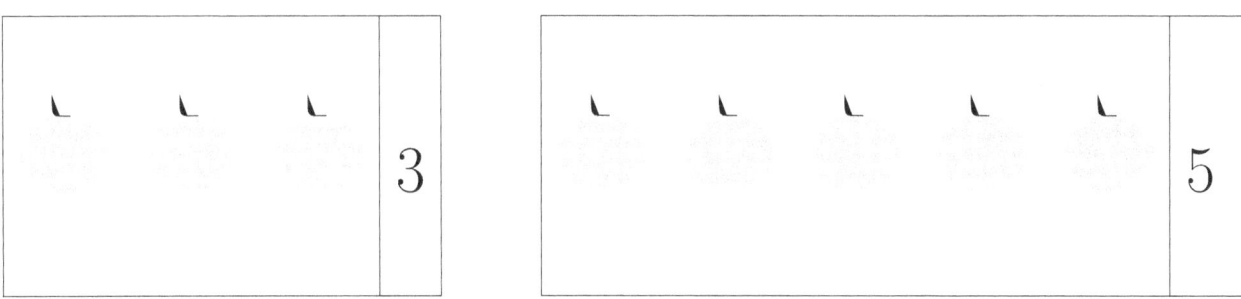

• In the left box there are less apples than in the right box.

We say 3 less than 5

or 3 smaller than 5.

We write $3 < 5$

• In the right box there are more apples than in the left box.

We say 5 more than 3  (or 5 greater than 3) (or 5 bigger than 3.)

We write $5 > 3$

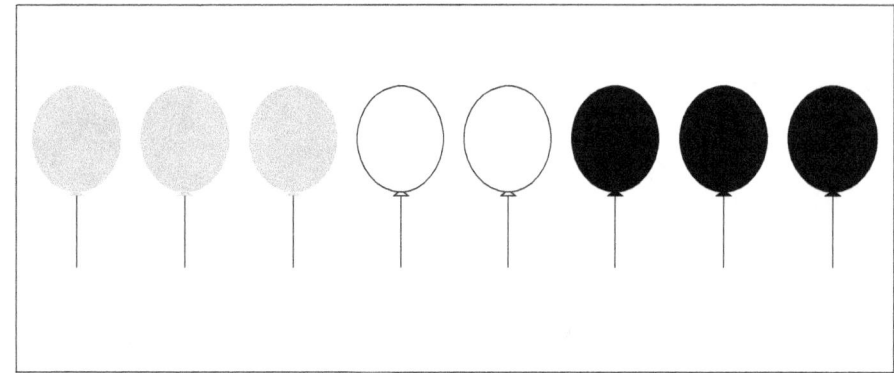

• There are more black balloons than white balloons

3 is greater than 2          3 > 2

• There are less white balloons than gray balloons

2 is less than 3          2 < 3

• There are the same number of black balloons as gray balloons

3 equal to 3          3 = 3

**Exercise 1: Fill in the blanks with more, less, >, < or =**

- There are ........... white circles than gray circles
- There are .......... gray circles than black circles
- 3 ... 9      - 9 ... 3      - 3 ... 3

**Exercise 2 : Fill in the blanks with more, less, > or <**

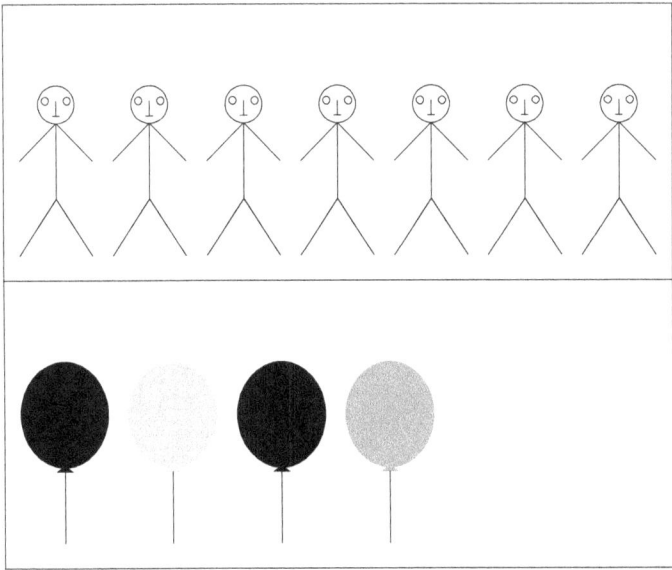

- There are .......... childrens than balloons
- There are .......... balloons than childrens
- 7 ... 4      - 4 ... 7

**Exercise 3 : Fill in the blanks with greater, less, > or <**

- Seven is .......... than four. So    7 ... 4.
- Four is .......... than seven . So    4 ... 7.
- Three is .......... than ten. So    3 ... 10.
- Ten is .......... than three . So    10 ... 3.
- One is .......... than five . So    1 ... 5.
- Six is .......... than eight. So    6 ... 8.
- Nine is .......... than two . So    9 ... 2.

**Exercise 4 : Fill in the blanks with > , < or =**

| 1 ... 2 | 4 ... 1 | 5 ... 9 |
| 7 ... 5 | 4 ... 7 | 6 ... 10 |
| 0 ... 5 | 7 ... 7 | 10 ... 8 |

# Ascending order and descending order

| • • • • • • • • • | • • | • • • • • |
|---|---|---|
| 9 | 2 | 5 |

- The least number is 2.

- The greatest number is 9.

- $2 < 5 < 9$. The ascending order is 2, 5, 9.

- $9 > 5 > 2$. The descending order is 9, 5, 2.

# Exercise 5

| • • • • • • • • • | • | • • • • • • • • • • | • • • • • • |
|---|---|---|---|
| 9 | 1 | 10 | 6 |

- The least number is ...
- The greatest number is ...
- $... < ... < ... < ....$
The ascending order is ..., ..., ..., ....
- $... > ... > ... > ... .$
The descending order is ..., ..., ..., ....

# Exercise 6 : Arrange the number in ascending order.

| 7 | 1 | 3 | 10 | 5 |

| | | | | |

| 2 | 8 | 4 | 10 | 6 |

| | | | | |

| 9 | 1 | 3 | 7 | 5 |

| | | | | |

| 5 | 3 | 5 | 1 | 2 |

| | | | | |

| 9 | 6 | 8 | 7 | 10 |

| | | | | |

| 4 | 6 | 5 | 7 | 8 |

| | | | | |

# Exercise 7 : Arrange the number in descending order.

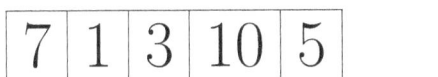  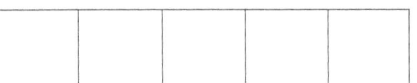

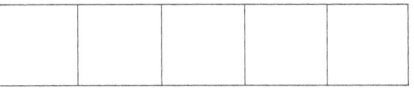

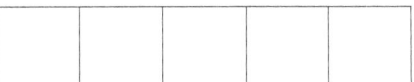

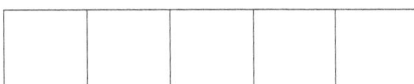

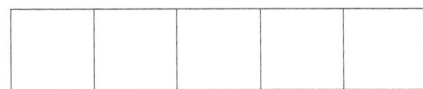

**Exercise 8 : Circle the least (smallest) number.**

| 7 | 1 | 9 | 10 | 5 |

| 8 | 6 | 9 | 10 | 5 |

**Exercise 9 : Circle the greatest (biggest) number.**

| 6 | 1 | 7 | 2 | 5 |

| 8 | 6 | 9 | 3 | 5 |

**Exercise 10 : Fill in the blanks with smallest, biggest, bigger or smaller, then give the orders.**

| 6 | 9 | 7 | 2 | 5 |

- 2 is the ................ number of the list.

- 2 is the .............. number of the list.

- 9 is the .............. number of the list.

- 2 is .............. than 7.

- 5 is .............. than 2.

- The ascending order is ..., ..., ..., ..., ....

- The descending order is ..., ..., ..., ...., ....

# Chapter 6

# Addition

- I buy 3 gray balloons and 2 white balloons.

How many balloons I have?

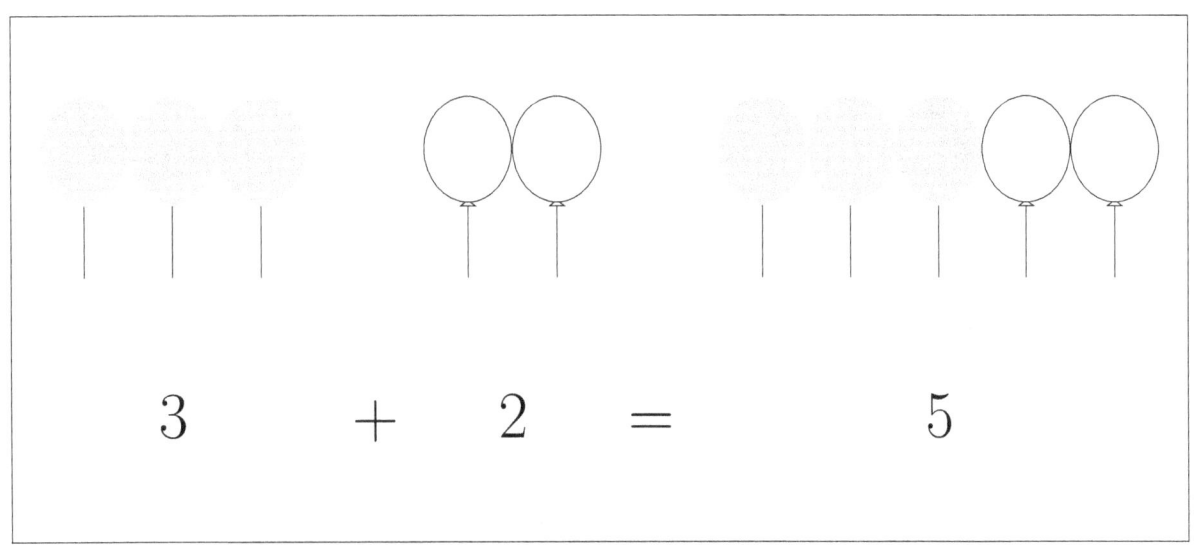

I have 5 balloons.

$$3 + 2 = 5$$

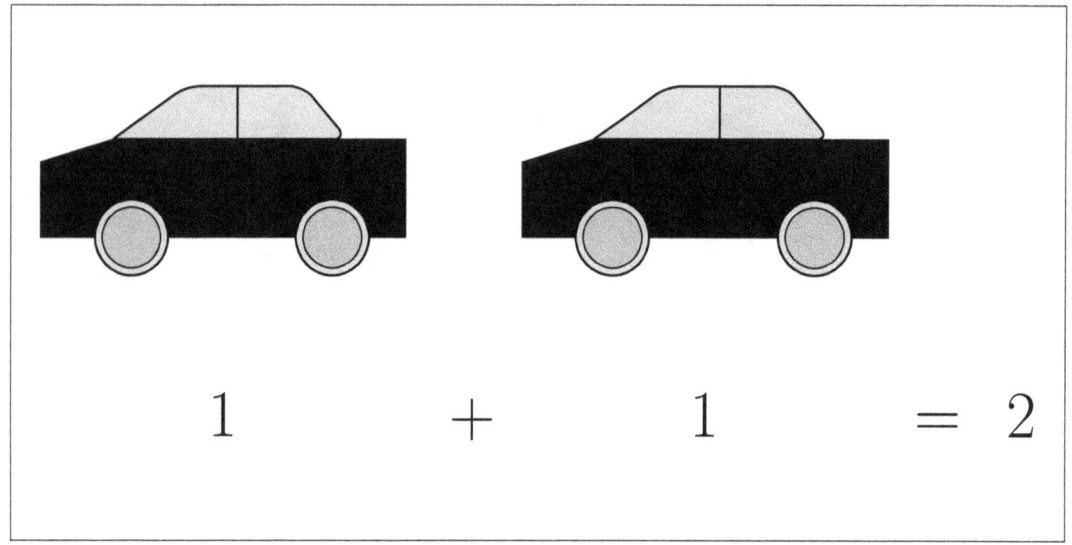

● My modher have one car and my fadher have one car.

How many cars we have ?

We have two cars.

$$1 + 1 = 2$$

$$0 + 0 = 0$$

$$5 + 0 = 5$$

Zero is the identity element for addition of numbers

# Exercise 1 : Sum up.

☆☆☆☆ ☆☆☆

4 + 3 =

△△△△△ △△△△

5 + 4 =

6 + 2 =

3 + 1 =

**Exercise 2 : Sum up.**

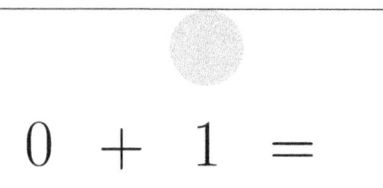

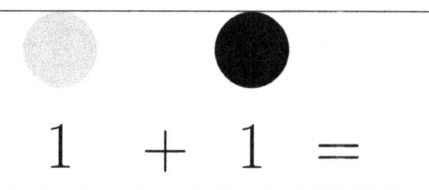

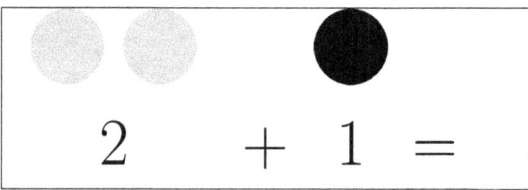

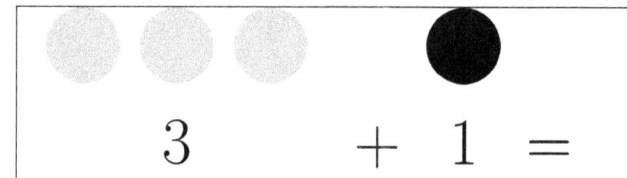

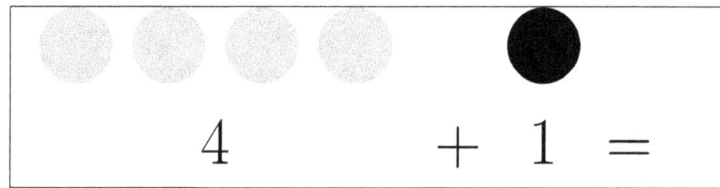

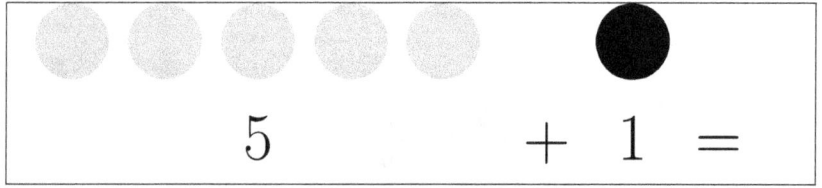

When we add a number whith 1, we get the next number.

$6 + 1 =$          ; $7 + 1 =$          ; $8 + 1 =$

$9 + 1 =$          ; $10 + 1 =$

# Exercise 3 : Sum up.

● ●     ● ●

2   +   2   =

● ● ●     ● ● ●

3   +   3   =

● ● ● ●     ● ● ● ●

4   +   4   =

● ● ● ● ●     ● ● ● ● ●

5   +   5   =

● ● ● ●     ● ● ● ● ●

4   +   5   =

● ● ● ● ●     ● ● ●

5   +   3   =

● ● ● ●     ● ● ●

4   +   3   =

●●●●●●      ●●●
$$6 \quad + \quad 3 \quad =$$

●●●●●●●      ●●
$$7 \quad + \quad 2 \quad =$$

●●●●●●      ●●
$$6 \quad + \quad 2 \quad =$$

●●●●●●●●      ●●
$$8 \quad + \quad 2 \quad =$$

●●●●●●      ●●●
$$6 \quad + \quad 3 \quad =$$

●●●●●●●      ●●●
$$7 \quad + \quad 3 \quad =$$

# Commutativity of addition

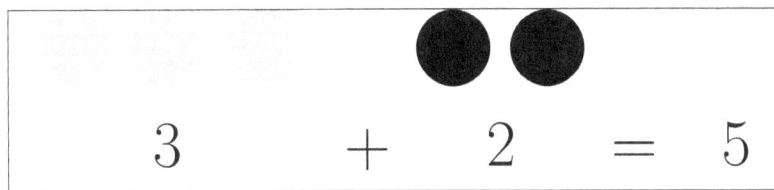

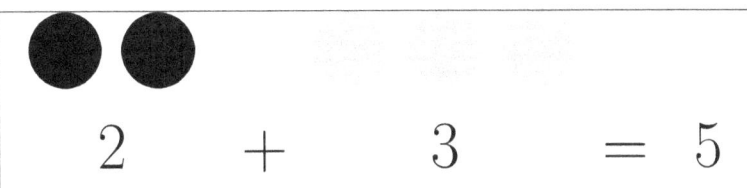

We have : 3+2=2+3=5


## Addition is commutative


**Exercise 4 : Complete the tables below.**

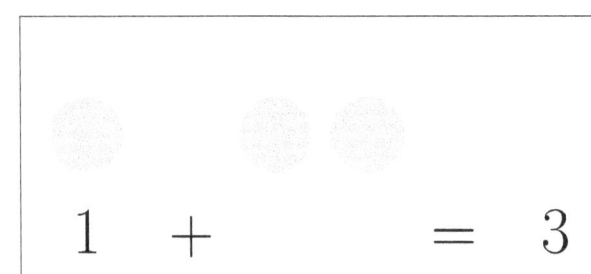

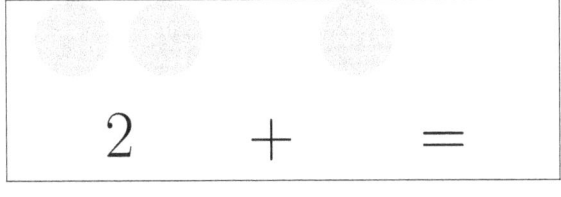

**Exercise 5 : Complete the tables below with the correct numbers.**

| 0 + = 4 | 4 + = 4 | 3 + = 4 |

1 + = 4

2 + =

**Exercise 6 : Complete the tables below with the correct numbers.**

| 0 + = 5 | 5 + = 5 |

1 + = 5

2 + =

| 3 + = 5 | 4 + = 5 |

**Exercise 7 :** Complete the tables below with the correct numbers.

$$0 \; + \quad = \quad 6$$

$$6 \; + \quad = \quad 6$$

$$1 \; + \qquad\qquad = \quad 6$$

$$2 \; + \qquad\qquad =$$

$$3 \quad + \qquad\qquad = \quad 6$$

$$5 \; + \quad = \quad 6$$

$$4 \; + \quad = \quad 6$$

**Exercise 8 :** **Complete the tables below with the correct numbers.**

$$0 \ + \ \ = \ 7 \qquad\qquad 7 \ + \ \ = \ 7$$

$$1 \ + \ \qquad\qquad\qquad = \ 7$$

$$2 \quad + \qquad\qquad\qquad =$$

$$3 \quad + \qquad\qquad\qquad = \ 7$$

$$6 \ + \ \ = \ 7 \qquad 5 \ + \ \ = \ 7 \qquad 4 \ + \ \ = \ 7$$

There are seven days in a week :
monday, tuesday, wednesday, thursday, friday,
saturday and sunday.

**Exercise 9 :** Complete the tables below with the correct numbers.

$0 + \quad = \quad 8$ $\qquad$ $8 + \quad = \quad 8$

$1 + \qquad\qquad\qquad = \quad 8$

$2 + \qquad\qquad\qquad =$

$3 + \qquad\qquad\qquad = \quad 8$

$4 + \qquad\qquad\qquad = \quad 8$

$7 + \quad = \quad 8$ $\qquad$ $6 + \quad = \quad 8$ $\qquad$ $5 + \quad = \quad 8$

**Exercise 10 : Complete the tables below with the correct numbers.**

| | |
|---|---|
| 0 + = 9 | 9 + = 9 |

$1 + \qquad = 9$

$2 + \qquad =$

$3 + \qquad = 9$

$4 + \qquad = 9$

| | |
|---|---|
| 8 + = 9 | 7 + = 9 |

| | |
|---|---|
| 6 + = 9 | 5 + = 9 |

**Exercise 11 : Complete the tables below with the correct numbers.**

$$0 \; + \quad = \quad 10 \qquad\qquad 10 \; + \quad = \quad 10$$

$$1 \; + \qquad\qquad\qquad\qquad = \quad 10$$

$$2 \; + \qquad\qquad\qquad\qquad =$$

$$3 \quad + \qquad\qquad\qquad = \quad 10$$

$$4 \quad + \qquad\qquad\qquad = \quad 10$$

$$5 \quad + \qquad\qquad\qquad = \quad 10$$

$$9 \; + \quad = \quad 10 \qquad\qquad 8 \; + \quad = \quad 10$$

$$7 \; + \quad = \quad 10 \qquad\qquad 6 \; + \quad = \quad 10$$

**Exercise 12 : Calculate the sums of numbers.**

| 5  +  1  = |

| 5  +  2  = |

| 5  +  3  = |

| 5  +  4  = |

**Calcul whith line numbers**

- 6+3=?

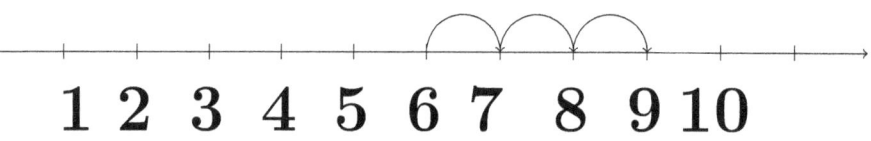

So 6+3=9

- 8+2=?

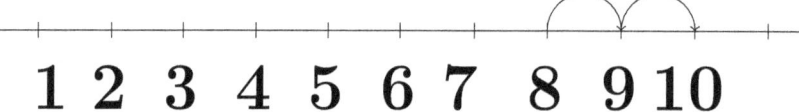

So 8+2=10

- 7+3=?

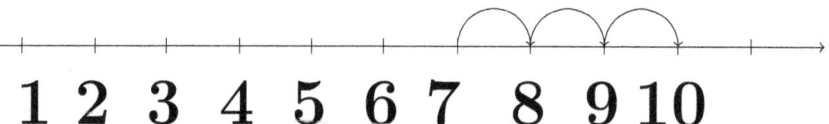

So 7+3=10

**Exercise 13 : Calculate the sum with line numbers.**

- 5+2=?

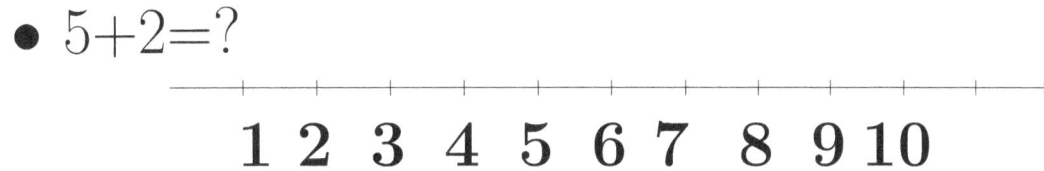

1 2 3 4 5 6 7 8 9 10

So 5+2=

- 5+1=?

1 2 3 4 5 6 7 8 9 10

So 5+1=

- 5+3=?

1 2 3 4 5 6 7 8 9 10

So 5+3=

- 5+4=?

1 2 3 4 5 6 7 8 9 10

So 5+4=

- 5+5=?

1 2 3 4 5 6 7 8 9 10

So 5+5=

**Exercise 14 : Calculate the sum with line numbers.**

- $7+2=?$

$$1 \quad 2 \quad 3 \quad 4 \quad 5 \quad 6 \quad 7 \quad 8 \quad 9 \quad 10$$

So $7+2=$

- $8+1=?$

$$1 \quad 2 \quad 3 \quad 4 \quad 5 \quad 6 \quad 7 \quad 8 \quad 9 \quad 10$$

So $8+1=$

- $3+3=?$

$$1 \quad 2 \quad 3 \quad 4 \quad 5 \quad 6 \quad 7 \quad 8 \quad 9 \quad 10$$

So $3+3=$

- $7+2=?$

$$1 \quad 2 \quad 3 \quad 4 \quad 5 \quad 6 \quad 7 \quad 8 \quad 9 \quad 10$$

So $7+2=$

- $4+3=?$

$$1 \quad 2 \quad 3 \quad 4 \quad 5 \quad 6 \quad 7 \quad 8 \quad 9 \quad 10$$

So $4+3=$

**Mental calcul.**

• 6+3= ? I make 6 in my minde, I tell the three next numbers, so the sum is the third number.

7 $\longrightarrow$ 8 $\longrightarrow$ 9. So $\boxed{6 \ + \ 3 \ = \ 9}$

• 2+8= ? I make 8 in my minde, I tell the two next numbers, so the sum is the second number.

9 $\longrightarrow$ 10. So $\boxed{2 \ + \ 8 \ = \ 10}$

**Exercise 15 : Calculate the sums of numbers.**

| | |
|---|---|
| $6 \ + \ 2 \ =$ | $7 \ + \ 3 \ =$ |
| $5 \ + \ 4 \ =$ | $5 \ + \ 2 \ =$ |
| $4 \ + \ 6 \ =$ | $0 \ + \ 7 \ =$ |
| $2 \ + \ 2 \ =$ | $1 \ + \ 1 \ =$ |
| $4 \ + \ 4 \ =$ | $2 \ + \ 7 \ =$ |
| $5 \ + \ 3 \ =$ | $3 \ + \ 2 \ =$ |

**Exercise 16 : Complete the missing number.**

| $+ \; 3 \; = \; 6$ | $+ \; 4 \; = \; 8$ |

| $+ \; 2 \; = \; 4$ | $+ \; 5 \; = \; 10$ |

| $+ \; 1 \; = \; 2$ | $+ \; 5 \; = \; 7$ |

| $+ \; 3 \; = \; 10$ | $+ \; 5 \; = \; 8$ |

| $+ \; 6 \; = \; 7$ | $+ \; 8 \; = \; 9$ |

| $+ \; 8 \; = \; 10$ | $+ \; 6 \; = \; 9$ |

| $+ \; 3 \; = \; 5$ | $+ \; 5 \; = \; 5$ |

**Vertical addition.**

$$
\begin{array}{r}
3 \\
+ \\
2 \\
\hline
= \; 5
\end{array}
\qquad
\begin{array}{r}
6 \\
+ \\
4 \\
\hline
= \; 10
\end{array}
$$

**Exercsce 17 : Find the sums below.**

$$2 + 2 =$$

$$3 + 3 =$$

$$4 + 4 =$$

$$5 + 5 =$$

$$4 + 3 =$$

$$4 + 2 =$$

$$2 + 5 =$$

$$3 + 6 =$$

$$4 + 1 =$$

$$5 + 4 =$$

$$7 + 3 =$$

$$8 + 2 =$$

**Exercise 18 : Complete the missing numbers.**

| + | 1 | 2 | 3 | 4 | 5 | 6 | 7 | 8 | 9 |
|---|---|---|---|---|---|---|---|---|---|
| 1 |   |   |   | 5 |   |   |   |   |   |

**Exercise 19 : Complete the missing numbers.**

| + | 1 | 2 | 3 | 4 | 5 | 6 | 7 | 8 |
|---|---|---|---|---|---|---|---|---|
| 2 |   |   | 5 |   |   |   |   |   |

**Exercise 20 : Complete the missing numbers.**

| + | 1 | 2 | 3 | 4 | 5 | 6 | 7 |
|---|---|---|---|---|---|---|---|
| 3 |   |   | 6 |   |   |   |   |

**Exercise 21 : Complete the missing numbers.**

| + | 1 | 2 | 3 | 4 | 5 | 6 |
|---|---|---|---|---|---|---|
| 4 |   |   | 7 |   |   |   |

**Exercise 22 : Complete the missing numbers.**

| + | 1 | 2 | 3 | 4 | 5 |
|---|---|---|---|---|---|
| 5 |   |   | 8 |   |   |

**Exercise 23 : Complete the missing numbers.**

| + | 1 | 2 | 3 | 4 | 5 |
|---|---|---|---|---|---|
| 1 |   |   |   |   |   |
| 2 |   |   |   |   |   |
| 3 |   |   |   |   |   |
| 4 |   | 6 |   |   |   |
| 5 |   |   |   |   |   |

**Exercise 24 : Complete the missing numbers.**

| + | 1 | 2 | 3 | 4 | 5 | 6 |
|---|---|---|---|---|---|---|
| 1 |   |   |   |   |   |   |
| 2 |   |   | 5 |   |   |   |
| 3 |   |   |   |   |   |   |
| 4 |   |   |   |   |   |   |

**Exercise 25 : Complete the missing numbers.**

| + | 1 | 2 | 3 | 4 | 5 | 6 | 7 |
|---|---|---|---|---|---|---|---|
| 1 |   |   |   |   |   |   |   |
| 2 |   |   | 5 |   |   |   |   |
| 3 |   |   |   |   |   |   |   |

- We decompose three as sum of two numbers

$$3 = 0 + 3 \qquad\qquad 3 = 3 + 0$$

$$3 = 1 + 2 \qquad\qquad 3 = 2 + 0$$

- We decompose zero as sum of two numbers

$$0 = 0 + 0$$

- We decompose one as sum of two numbers

$$1 = 0 + 1 \qquad\qquad 1 = 1 + 0$$

- We decompose two as sum of two numbers.

$$2 = 0 + 2 \qquad 2 = 2 + 0 \qquad 2 = 1 + 1$$

**Exercise 26 : Decompose four as sum of two numbers.**

$$4 = \quad + \qquad 4 = \quad + \qquad 4 = \quad +$$

$$4 = \quad + \qquad 4 = \quad +$$

**Exercise 27 : Decompose five as sum of two numbers.**

5 = +      5 = +      5 = +

5 = +      5 = +      5 = +

**Exercise 28 : Decompose six as sum of two numbers.**

6 = +      6 = +      6 = +

6 = +      6 = +      6 = +

6 = +

**Exercise 29 : Decompose seven as sum of two numbers.**

7 = +      7 = +      7 = +

7 = +      7 = +      7 = +

7 = +      7 = +

**Exercise 30 : Decompose eight as sum of two numbers.**

| | | |
|---|---|---|
| 8 = + | 8 = + | 8 = + |
| 8 = + | 8 = + | 8 = + |
| 8 = + | 8 = + | 8 = + |

**Exercise 31 : Decompose nine as sum of two numbers.**

| | | |
|---|---|---|
| 9 = + | 9 = + | 9 = + |
| 9 = + | 9 = + | 9 = + |
| 9 = + | 9 = + | 9 = + |
| 9 = + | | |

**Exercise 32 : Decompose ten as sum of two numbers.**

| | | |
|---|---|---|
| 10 = + | 10 = + | 10 = + |
| 10 = + | 10 = + | 10 = + |
| 10 = + | 10 = + | 10 = + |
| 10 = + | 10 = + | |

# Exercise 33 : Complete the missing numbers.

□ + □ = 2
+     +
□ + □ = 9
=     =
5     6

□ + □ = 1
+     +
□ + □ = 10
=     =
7     4

□ + □ = 1
+     +
□ + □ = 9
=     =
5     5

□ + □ = 0
+     +
□ + □ = 10
=     =
6     4

□ + □ = 4
+     +
□ + □ = 8
=     =
4     8

□ + □ = 3
+     +
□ + □ = 9
=     =
7     5

**Exercise 33 : Sum up.**

$$1 + 1 + 1 =$$

$$2 + 2 + 2 =$$

$$3 + 3 + 3 =$$

$$5 + 3 + 2 =$$

$$4 + 4 + 2 =$$

$$4 + 2 + 2 =$$

$$5 + 2 + 2 =$$

**Exercise 34 : Sum up. You can represent the sum with dots.**

$2 + 2 + 1 =$

$7 + 2 + 1 =$

$5 + 2 + 1 =$

$2 + 1 + 1 =$

$2 + 2 + 2 =$

$3 + 3 + 3 =$

$3 + 3 + 1 =$

$2 + 3 + 5 =$

$4 + 2 + 4 =$

$2 + 2 + 2 + 2 =$

$4 + 2 + 2 =$

$$4 + 2 + 1 + 1 =$$

## Exercise 35 : Write the missing numbers.

$$\boxed{\quad + 2 + 2 = 6}$$

$$\boxed{5 + 2 + \quad = 8}$$

$$\boxed{3 + \quad + 3 = 9}$$

$$\boxed{2 + 2 + 2 + \quad = 8}$$

$$\boxed{2 + 2 + 2 + 2 + \quad = 10}$$

$$\boxed{4 + 4 + \quad = 8}$$

$$\boxed{4 + 4 + \quad = 10}$$

$$\boxed{2 + 2 + \quad = 7}$$

$$\boxed{2 + 2 + \quad = 8}$$

$$\boxed{2 + 2 + 2 + \quad = 10}$$

**Exercise 36**

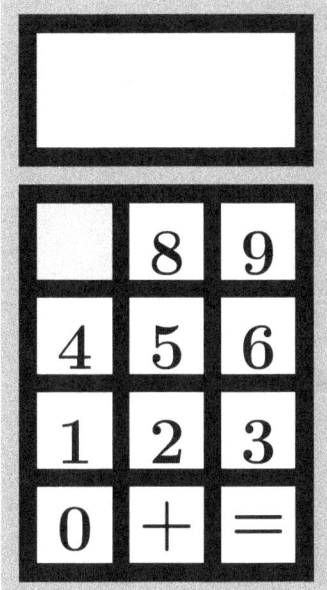

The number 7 key on the calculator is broken.

How we can make the number 7 appear on the screen without the 7 key? (Give four possibilities)

.................................................................................

.................................................................................

.................................................................................

.................................................................................

**Exercise 37**

Which key number is broken on the calculator ?

..........................................................................

How we can make the number 10 appear on the screen without the 1 key ? (Give three possibilities)

..........................................................................

..........................................................................

..........................................................................

**Exercise 38**

How many keys are broken on the calculator?

..................................................................

Which keys numbers are broken ?

..................................................................

How we can make the number 9 appear on the screen without the 0 key, 3 key, 4 key, 6 key, 7 key, 8 key and 9 key ? (With minimal touch)

..................................................................

**Exercise 39**

How we can make the number 4 appear on the screen of this calculator? (With minimal touch)

........................................................................

How we can make the number 8 appear on the screen of this calculator? (With minimal touch)

........................................................................

How we can make the number 10 appear on the screen of this calculator? (With minimal touch)

........................................................................

**Exercise 40**

How we can make the number 6 appear on the screen of this calculator? (With minimal touch)

..................................................................

How we can make the number 9 appear on the screen of this calculator? (With minimal touch)

..................................................................

How we can make the number 10 appear on the screen of this calculator? (With minimal touch)

..................................................................

**Exercise 41**

5 ducks in the lake.

4 ducks arrive at the lake.

How many ducks are there?

......................................................

**Exercise 42**

There 4 bees in garden.

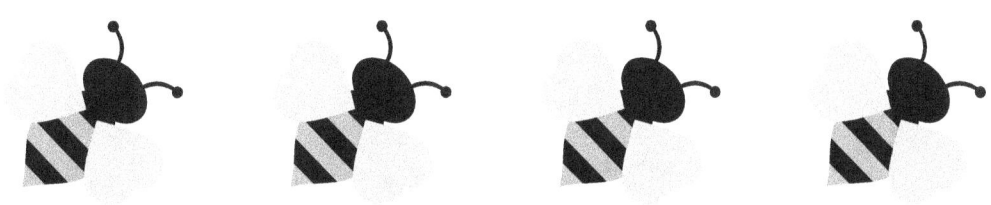

4 bees arrive at garden.

How many bees are there?

.....................................................................

# Chapter 7

# Subtraction

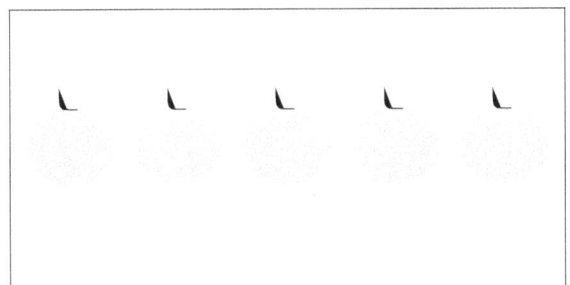

I have 5 apples. I eat 2 apples.

How many apples left?

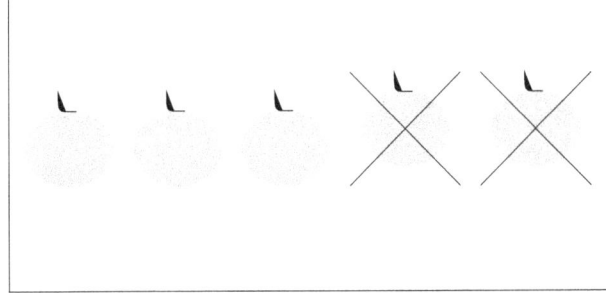

3 apples left.

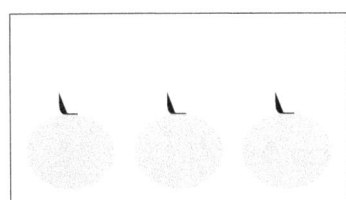

$$5 - 2 = 3$$

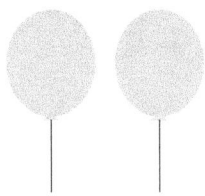

I have 2 balloons. One balloon brust.

How many balloons left?

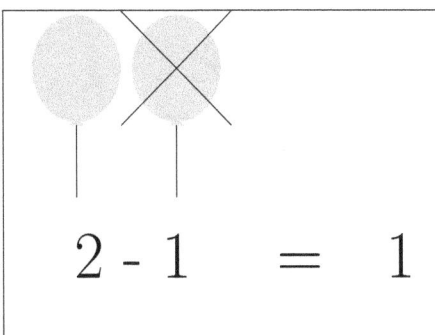

One balloon left.

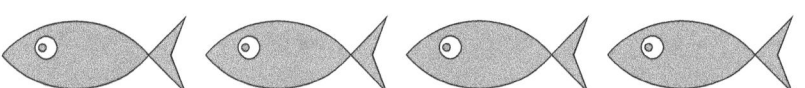

I have 4 fish. The cat eat 2 fish.

How many fish left?

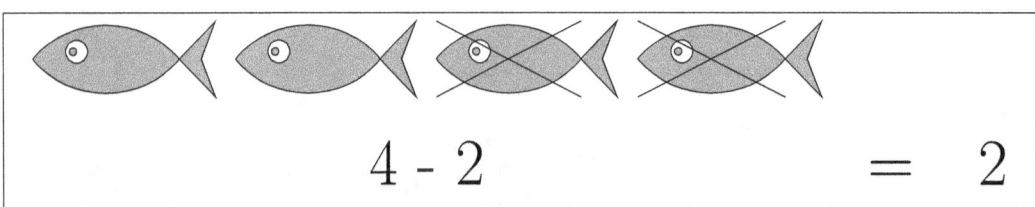

2 fish left.

# Exercise 1 : Finde the difference.

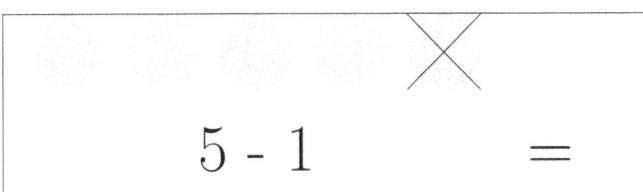

5 - 1          =

8 - 2               =

10 - 3               =

7 - 4          =

9 - 5               =

8 - 5               =

**Exercise 2 : Finde the difference.**

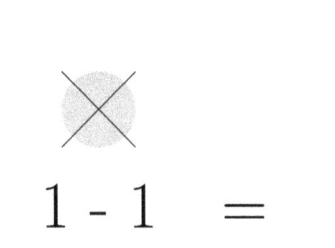

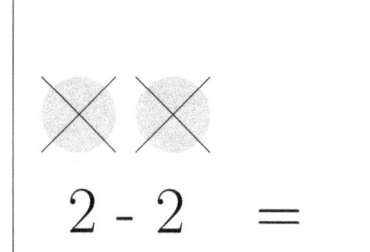

| | |
|---|---|
| 3 - 3   = | 4 - 4        = |

5 - 5          =

When we substract a number from itself we get zero : 10 - 10 = 0

6 - 6   =          7 - 7   =

8 - 8   =          9 - 9   =

**Exercice 3 : Finde the difference.**

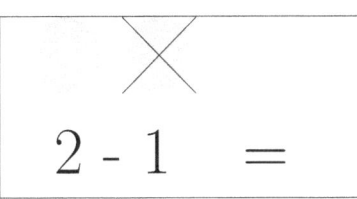

2 - 1   =

3 - 1    =

4 - 1     =

5 - 1     =

When we substract 1 from a number we get the predecessor of this number : 10 - 1 = 9

6 - 1   =          7 - 1   =

8 - 1   =          9 - 1   =

# Exercise 4 : Finde the numbers missing

• • • • ✕
$$5 - \quad = \quad 4$$

• • • • • • ✕ ✕
$$\quad - 2 \quad = \quad 6$$

• • • • • ✕ ✕ ✕
$$8 - \quad = \quad$$

• • • • • ✕ ✕ ✕
$$\quad - \quad = \quad$$

• • • ✕ ✕ ✕ ✕ ✕
$$\quad - \quad = \quad$$

• • • • • ✕ ✕ ✕
$$\quad - \quad = \quad$$

• • • • ✕ ✕ ✕
$$\quad - \quad = \quad$$

• • • • ✕ ✕
$$\quad - \quad = \quad$$

• • • • • • ✕ ✕
$$\quad - \quad = \quad$$

• • • ✕ ✕ ✕
$$\quad - \quad = \quad$$

• • • • • • ✕ ✕
$$\quad - \quad = \quad$$

• • ✕ ✕ ✕
$$\quad - \quad = \quad$$

• • • • ✕ ✕ ✕ ✕ ✕ ✕
$$\quad - \quad = \quad$$

• • • • • • ✕ ✕
$$\quad - \quad = \quad$$

107

# Exercise 5 : Cross off the circles and find the difference

5 - 2 =

8 - 4 =

10 - 6 =

7 - 0 =

9 - 3 =

8 - 4 =

# Exercise 6 : Cross off the circles and find the difference

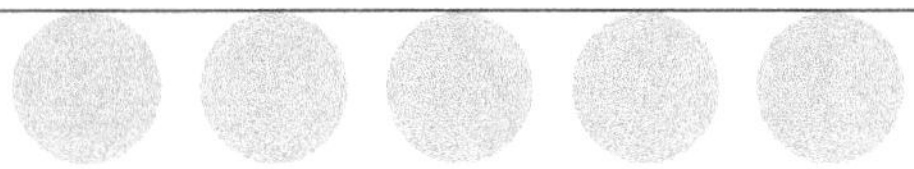

5 - 4        =

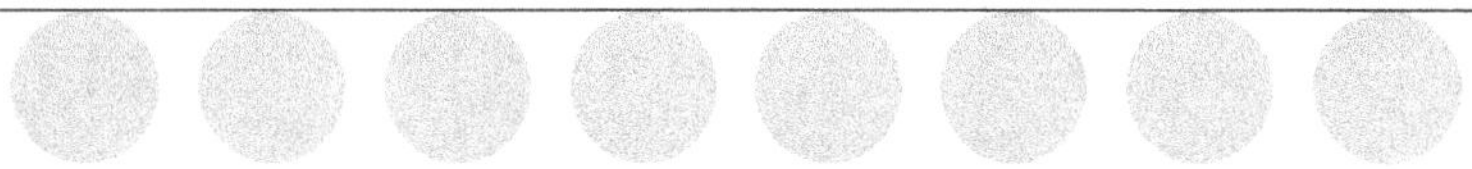

8 - 2                =

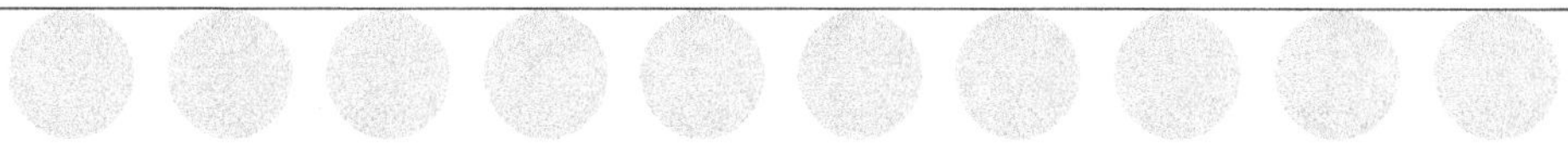

10 -5                      =

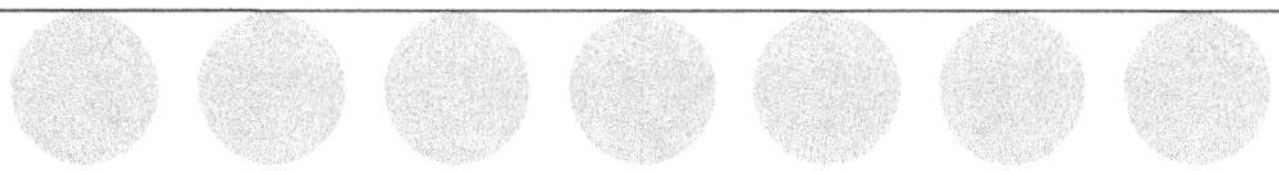

7 - 2            =

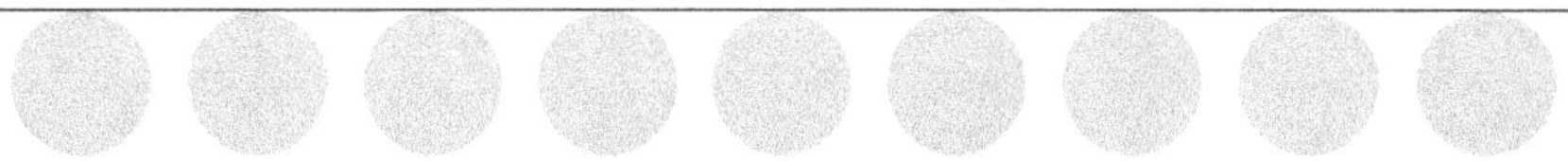

9 - 4                  =

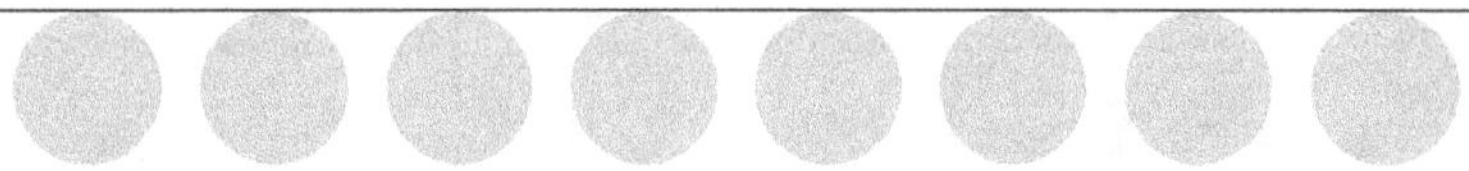

8 - 2              =

# Exercise 7 : Cross off the circles and find the difference

7 - 3          =

8 - 8          =

10 -2          =

7 - 1          =

9 -8          =

8 - 3          =

**Exercise 8 : Draw the dots missing and cross od to finde the difference**

5 - 3     =

7 - 3        =

5 - 2     =

7 - 1        =

6 - 3     =

10 - 2         =

8 - 4        =

7 - 2        =

6 - 4     =

10 - 5         =

6 - 1     =

10 - 4         =

7 - 4        =

9 - 4          =

# Calcul whith line numbers

- 9-3=?

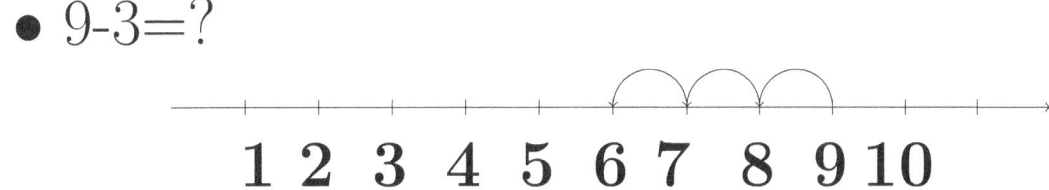

So 9-3=6

- 10-1=?

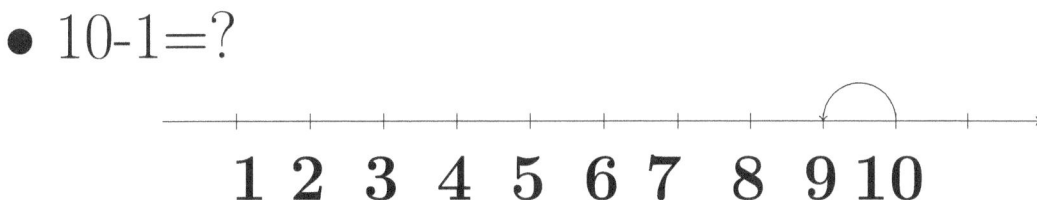

So 10-1=9

- 10-2=?

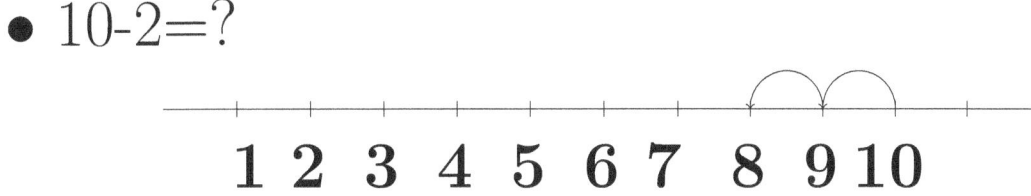

So 10-2=8

**Exercise 9 : Calculate the sum with line numbers.**

- 5-1=?

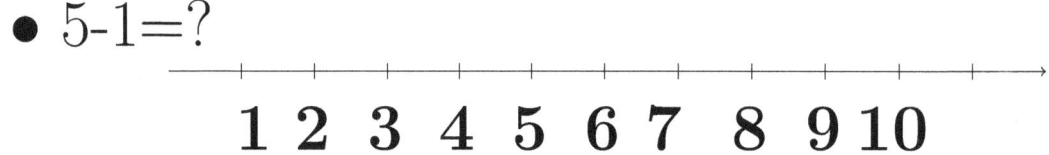

So 5-1=

- 5-2=?

1 2 3 4 5 6 7 8 9 10

So 5-2=

- 5-3=?

1 2 3 4 5 6 7 8 9 10

So 5-3=

- 5-4=?

1 2 3 4 5 6 7 8 9 10

So 5-4=

- 6-5=?

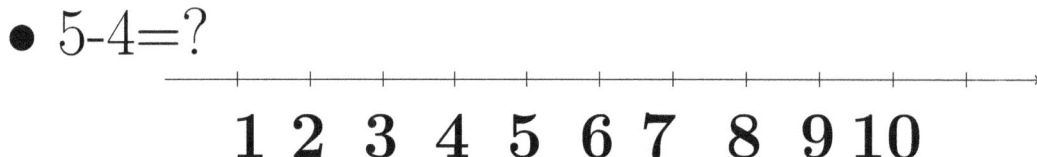

So 6-5=

**Exercise 10 : Calculate the sum with line numbers.**

- 7-2=?

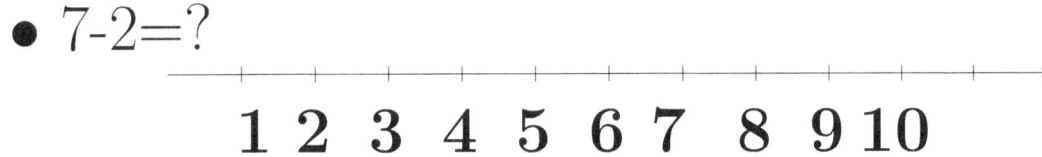

So 7-2=

- 8-3=?

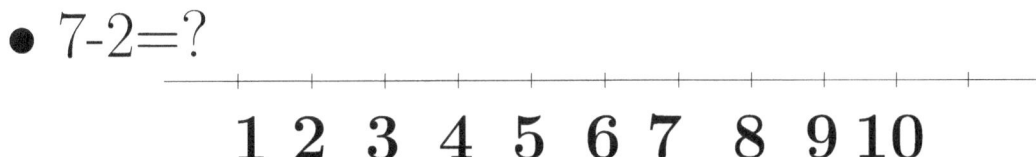

So 8-3=

- 10-6=?

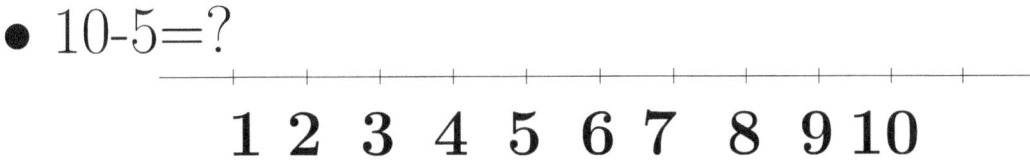

So 10-6=

- 7-2=?

So 7-2=

- 10-5=?

So 10-5=

**Addition and substraction are inverse operations.**

$$3 + 2 = 5$$

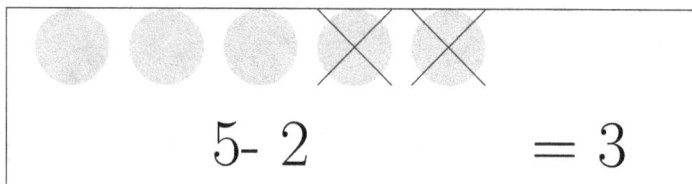

$$5 - 2 = 3$$

**Exercise 11 : Complet the missing number.**

$2 + \boxed{\phantom{0}} = 3$   so   $3 - 2 = \boxed{\phantom{0}}$

$5 + \boxed{\phantom{0}} = 6$   so   $6 - 5 = \boxed{\phantom{0}}$

$8 + \boxed{\phantom{0}} = 10$   so   $10 - 8 = \boxed{\phantom{0}}$

$7 + \boxed{\phantom{0}} = 10$   so   $10 - 7 = \boxed{\phantom{0}}$

$6 + \boxed{\phantom{0}} = 10$   so   $10 - 6 = \boxed{\phantom{0}}$

$5 + \boxed{\phantom{0}} = 9$   so   $9 - 5 = \boxed{\phantom{0}}$

$4 + \boxed{\phantom{0}} = 8$   so   $8 - 4 = \boxed{\phantom{0}}$

$5 + \boxed{\phantom{0}} = 7$   so   $7 - 5 = \boxed{\phantom{0}}$

**Exercise 12 : Finde the difference.**

$2 - 1 \quad =$

$4 - 2 \quad =$

$6 - 3 \quad =$

$8 - 4 \quad =$

$10 - 10 \quad =$

$0 - 0 \quad =$

$4 - 1 \quad =$

$9 - 7 \quad =$

$6 - 5 \quad =$

$7 - 3 \quad =$

$9 - 4 \quad =$

$7 - 5 \quad =$

$5 - 5 \quad =$

$9 - 8 \quad =$

$10 - 3 \quad =$

$6 - 2 \quad =$

# Vertical substraction.

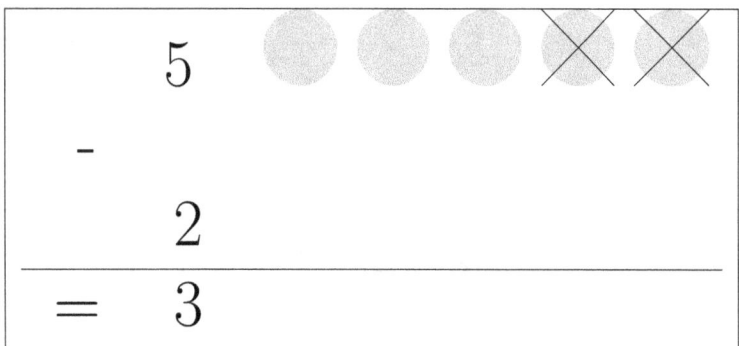

$$
\begin{array}{r}
5 \\
- \quad \\
2 \\
\hline
= \quad 3
\end{array}
$$

## Exercise 13: Calculate the sums below.

$$
\begin{array}{r}
2 \\
- \\
2 \\
\hline
=
\end{array}
\qquad
\begin{array}{r}
3 \\
- \\
2 \\
\hline
=
\end{array}
\qquad
\begin{array}{r}
4 \\
- \\
2 \\
\hline
=
\end{array}
$$

$$
\begin{array}{r}
5 \\
- \\
3 \\
\hline
=
\end{array}
\qquad
\begin{array}{r}
4 \\
- \\
1 \\
\hline
=
\end{array}
\qquad
\begin{array}{r}
6 \\
- \\
2 \\
\hline
=
\end{array}
$$

| | | |
|---|---|---|
| 10<br>+<br>9<br>――――<br>= | 6<br>-<br>3<br>――――<br>= | 8<br>-<br>4<br>――――<br>= |
| 10<br>-<br>5<br>――――<br>= | 7<br>-<br>3<br>――――<br>= | 8<br>-<br>2<br>――――<br>= |
| 10<br>-<br>7<br>――――<br>= | 10<br>-<br>6<br>――――<br>= | 4<br>-<br>0<br>――――<br>= |
| 9<br>-<br>7<br>――――<br>= | 8<br>-<br>6<br>――――<br>= | 9<br>-<br>5<br>――――<br>= |

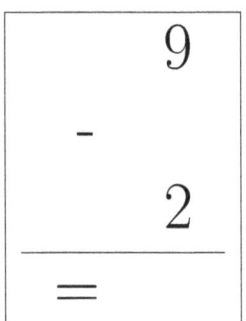
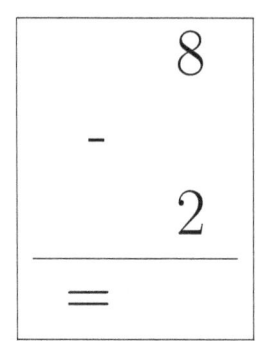
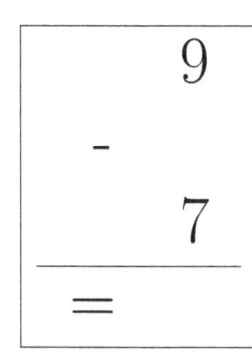

**Exercise 14 : Complete the missing numbers.**

| - | 1 | 2 | 3 | 4 | 5 | 6 | 7 | 8 | 9 | 10 |
|---|---|---|---|---|---|---|---|---|---|----|
| 1 |   | 2 |   |   |   |   |   |   |   |    |

**Exercise 15 : Complete the missing numbers.**

| - | 2 | 3 | 4 | 5 | 6 | 7 | 8 | 9 | 10 |
|---|---|---|---|---|---|---|---|---|----|
| 2 |   |   |   |   |   |   |   |   |    |

**Exercise 16 : Complete the missing numbers.**

| - | 3 | 4 | 5 | 6 | 7 | 8 | 9 | 10 |
|---|---|---|---|---|---|---|---|----|
| 3 |   |   |   |   |   |   |   |    |

**Exercise 17 : Complete the missing numbers.**

| - | 4 | 5 | 6 | 7 | 8 | 9 | 10 |
|---|---|---|---|---|---|---|----|
| 4 |   |   |   |   |   |   |    |

# Exercise 18 : Complete the missing numbers.

| -  | 5 | 6 | 7 | 8 | 9 | 10 |
|----|---|---|---|---|---|----|
| 1  |   |   |   |   |   |    |
| 2  |   |   |   |   |   |    |
| 3  |   |   |   |   |   |    |
| 4  |   |   |   |   |   |    |
| 5  |   |   |   |   |   | 5  |

# Exercise 19 : Complete the missing numbers.

| -  | 4 | 5 | 6 | 7 | 8 | 9 | 10 |
|----|---|---|---|---|---|---|----|
| 1  |   |   |   |   |   |   |    |
| 2  |   |   |   |   |   |   |    |
| 3  |   |   |   |   |   |   |    |
| 4  |   |   |   |   |   |   | 6  |

**Exercise 20**

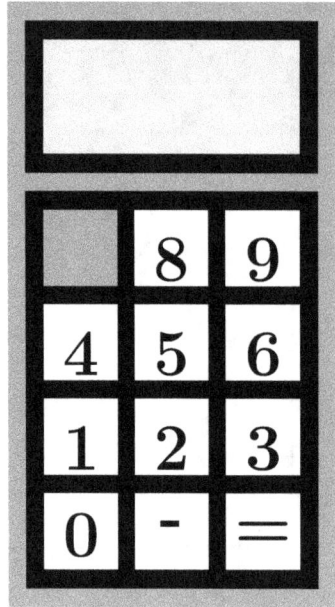

The number 7 key on the calculator is broken.

How we can make the number 7 appear on the screen without the 7 key? (Give three possibilities)

..................................................................

..................................................................

..................................................................

**Exercise 21**

Which key number is broken on the calculator ?

..................................................................................

How we can make the number 1 appear on the screen without the 1 key ? (Give three possibilities)

..................................................................................

..................................................................................

..................................................................................

**Exercise 22**

There are 10 ducks in the lake.

3 ducks go to the jungle.

How many ducks left in the lake?

..................................................................................

**Exercise 23**

There 9 bees in garden.

6 bees flew away.

How many bees left in the garden?

......................................................................................

**Exercise 24**

We need 8 tomatoes to make our sauce for dinner.

We have only 2 tomatoes.

How many more tomatoes do we need to buy ?

.................................................................................

.................................................................................

**Exercise 25 :**

Jhon is 10 years old, Sam is 5 years old and Jane is 3 years old.

What's the age difference between of Sam and Jane ?

..................................................................

What's the age difference between of Jhon and Sam ?

..................................................................

What's the age difference between of Jhon and Jane ?

..................................................................

**Exercise 26**

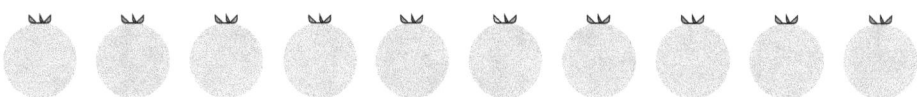

We need 10 tomatoes to make our sauce for dinner.

We have only 6 tomatoes.

How many more tomatoes do we need to buy ?

..................................................................................

..................................................................................

# Chapter 8

# Pattern with number

Exercise 1 : Complete the pattern.

| 2 | 4 | 6 | 8 | |
|---|---|---|---|---|

| 1 | 3 | 5 | 7 | |
|---|---|---|---|---|

| 6 | 7 | 8 | 9 | |
|---|---|---|---|---|

| 5 | 6 | 7 | 8 | |
|---|---|---|---|---|

| 4 | 5 | 6 | 7 | |
|---|---|---|---|---|

| 3 | 4 | 5 | 6 | |
|---|---|---|---|---|

# Exercise 2 : Complete the pattern.

| 1 | 1 | 2 | 1 | 1 | 2 |  |  |  |  |  |  |  |  |  |
|---|---|---|---|---|---|---|---|---|---|---|---|---|---|---|

| 2 | 2 | 4 | 2 | 2 | 4 |  |  |  |  |  |  |  |  |  |
|---|---|---|---|---|---|---|---|---|---|---|---|---|---|---|

| 3 | 3 | 6 | 3 | 3 | 6 |  |  |  |  |  |  |  |  |  |
|---|---|---|---|---|---|---|---|---|---|---|---|---|---|---|

| 4 | 4 | 8 | 4 | 4 | 8 |  |  |  |  |  |  |  |  |  |
|---|---|---|---|---|---|---|---|---|---|---|---|---|---|---|

| 5 | 5 | 10 | 5 | 5 | 10 |  |  |  |  |  |  |  | 5 |  |
|---|---|---|---|---|---|---|---|---|---|---|---|---|---|---|

| 4 | 2 | 2 | 4 | 2 | 2 |  |  |  |  |  |  |  |  |  |
|---|---|---|---|---|---|---|---|---|---|---|---|---|---|---|

| 6 | 3 | 3 | 6 | 3 | 3 |  |  |  |  |  |  |  |  |  |
|---|---|---|---|---|---|---|---|---|---|---|---|---|---|---|

| 8 | 4 | 4 | 8 | 4 | 4 |  |  |  |  |  |  |  |  |  |
|---|---|---|---|---|---|---|---|---|---|---|---|---|---|---|

| 10 | 5 | 5 | 10 | 5 | 5 |  |  |  |  |  |  |  |  |  |
|---|---|---|---|---|---|---|---|---|---|---|---|---|---|---|

# Exercise 3 : Complete the pattern.

| 1 | 1 | 1 | 3 | 1 | 1 | 1 | 3 |  |  |  |  |  |  |  |  |
|---|---|---|---|---|---|---|---|---|---|---|---|---|---|---|---|

| 2 | 2 | 2 | 6 | 2 | 2 | 2 | 6 |  |  |  |  |  |  |  |  |
|---|---|---|---|---|---|---|---|---|---|---|---|---|---|---|---|

| 3 | 3 | 3 | 9 | 3 | 3 | 3 | 9 |  |  |  |  |  |  |  |  |
|---|---|---|---|---|---|---|---|---|---|---|---|---|---|---|---|

| 2 | 2 | 2 | 2 | 8 | 2 | 2 | 2 | 2 | 8 |  |  |  |  |  |  |
|---|---|---|---|---|---|---|---|---|---|---|---|---|---|---|---|

| 2 | 2 | 2 | 2 | 2 | 10 | 2 | 2 | 2 | 2 | 2 |  |  |  |  |  |
|---|---|---|---|---|---|---|---|---|---|---|---|---|---|---|---|

| 3 | 1 | 1 | 1 | 3 | 1 | 1 | 1 |  |  |  |  |  |  |  |  |
|---|---|---|---|---|---|---|---|---|---|---|---|---|---|---|---|

| 6 | 2 | 2 | 2 | 6 | 2 | 2 | 2 |  |  |  |  |  |  |  |  |
|---|---|---|---|---|---|---|---|---|---|---|---|---|---|---|---|

| 9 | 3 | 3 | 3 | 9 | 3 | 3 | 3 |  |  |  |  |  |  |  |  |
|---|---|---|---|---|---|---|---|---|---|---|---|---|---|---|---|

| 8 | 2 | 2 | 2 | 2 | 8 | 2 | 2 | 2 | 2 |  |  |  |  |  |  |
|---|---|---|---|---|---|---|---|---|---|---|---|---|---|---|---|

| 10 | 2 | 2 | 2 | 2 | 2 | 10 |  |  |  | 10 |  |  |
|---|---|---|---|---|---|---|---|---|---|---|---|---|

**Exercise 4 : Complete the pattern.**

| 2 | 3 | 5 | 2 | 3 | 5 |  |  |  |  |  |  |  |  |
|---|---|---|---|---|---|---|---|---|---|---|---|---|---|

| 4 | 3 | 7 | 4 | 3 | 7 |  |  |  |  |  |  |  |  |
|---|---|---|---|---|---|---|---|---|---|---|---|---|---|

| 3 | 6 | 9 | 3 | 6 | 9 |  |  |  |  |  |  |  |  |
|---|---|---|---|---|---|---|---|---|---|---|---|---|---|

| 4 | 5 | 9 | 4 | 5 | 9 |  |  |  |  |  |  |  |  |
|---|---|---|---|---|---|---|---|---|---|---|---|---|---|

| 4 | 6 | 10 | 4 | 6 | 10 |  |  |  |  |  |  |  |  |
|---|---|---|---|---|---|---|---|---|---|---|---|---|---|

| 3 | 7 | 10 | 3 | 7 | 10 |  |  |  |  |  |  |  |  |
|---|---|---|---|---|---|---|---|---|---|---|---|---|---|

| 2 | 8 | 10 | 2 | 8 | 10 |  |  |  |  |  |  |  |  |
|---|---|---|---|---|---|---|---|---|---|---|---|---|---|

| 2 | 5 | 7 | 2 | 5 | 7 |  |  |  |  |  |  |  |  |
|---|---|---|---|---|---|---|---|---|---|---|---|---|---|

# Chapter 9

# Conting up to twenty

## 9.1   Eleven and twelve

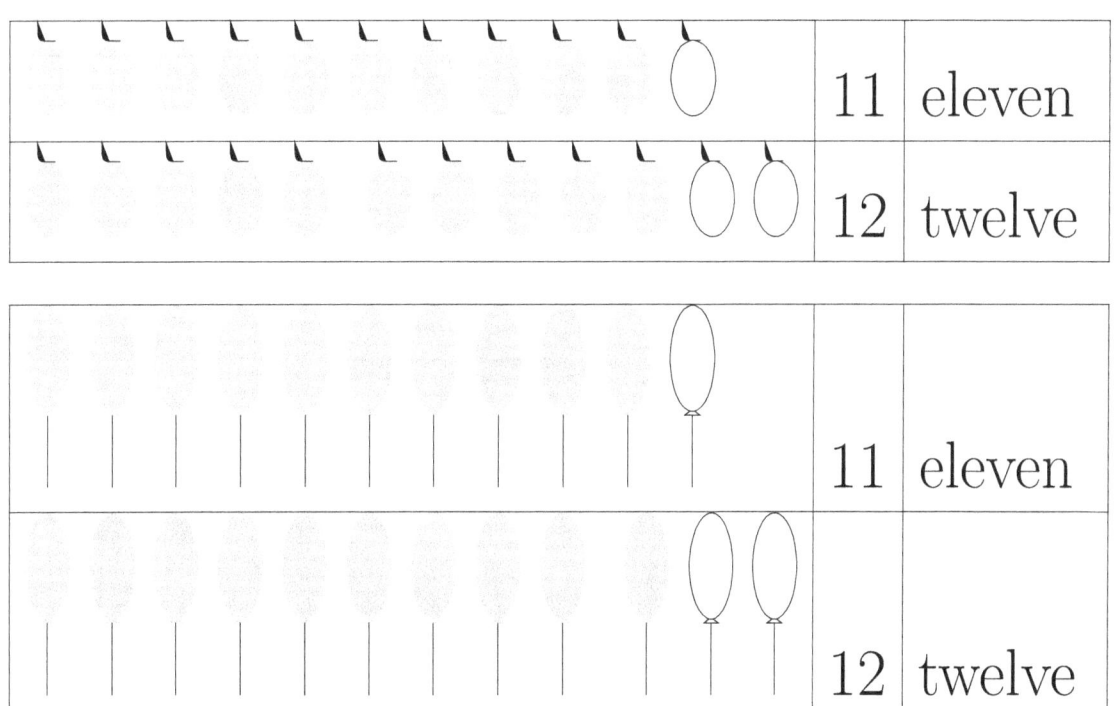

| | | 11 | eleven |
|---|---|---|---|
| | | 12 | twelve |

| | | 11 | eleven |
|---|---|---|---|
| | | 12 | twelve |

**Exercise 1 : Trace the number using a pencil or pen.**

11

12    12    12    12    12    12

Exercise 2 : How many circles are there ?

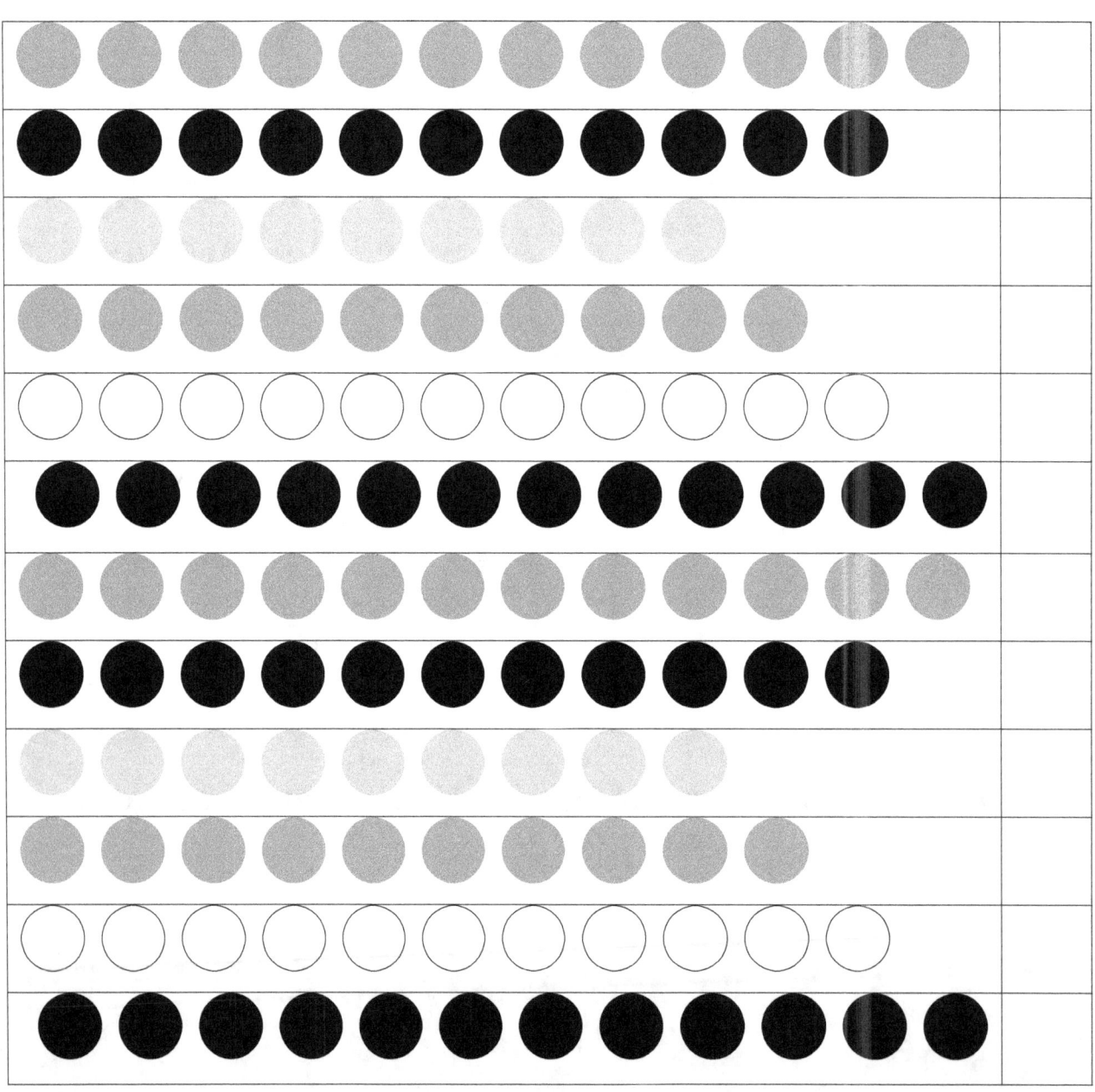

## 9.2 Thirteen and Fourteen

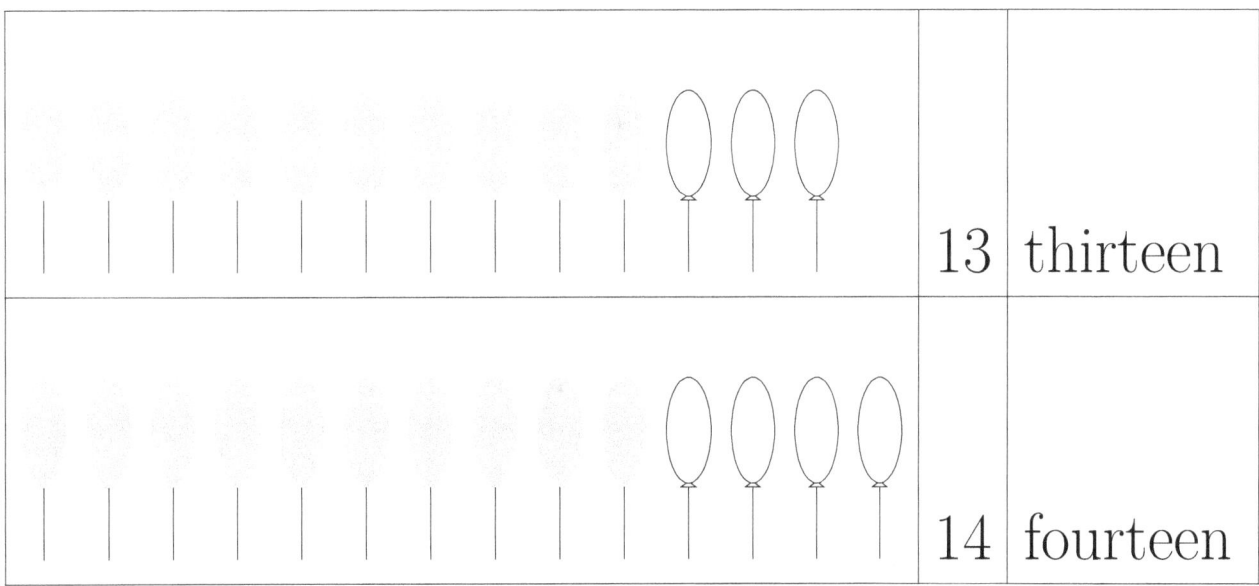

| | 13 | thirteen |
| --- | --- | --- |
| | 14 | fourteen |

**Exercise 3 : Trace the number using a pencil or pen.**

13

14

## 9.3    From fifteen to twenty

| ●●●●●●●●●●<br>●●●●● | 15 | fifteen |
|---|---|---|
| ●●●●●●●●●●<br>●●●●●● | 16 | sixteen |
| ●●●●●●●●●●<br>●●●●●●● | 17 | seventeen |
| ●●●●●●●●●●<br>●●●●●●●● | 18 | eighteen |
| ●●●●●●●●●●<br>●●●●●●●●● | 19 | nineteen |
| ●●●●●●●●●●<br>●●●●●●●●●● | 20 | twenty |

## Exercise 4 : Trace the number using a pencil or pen.

15

16

17

18

19

20

## Exercise 5: Match.

| | |
|---|---|
| 1 | four |
| 2 | three |
| 3 | two |
| 4 | one |
| 5 | seven |
| 6 | six |
| 7 | five |
| 8 | thrirteen |
| 9 | eleven |
| 10 | twelve |
| 11 | twenty |
| 12 | nineteen |
| 13 | eighteen |
| 14 | sixteen |
| 15 | nine |
| 16 | ten |
| 17 | fourteen |
| 18 | eight |
| 19 | seventeen |
| 20 | fifteen |

**Exercise 6**

There are 12 months in a year :

january, february, march, april, may, june, july, august, september, october, november, december.

The first month of year is january.

The second month of year is february.

The third month of year is ................

The fourth month of year is ................

The fifth month of year is ................

The sixth month of year is ................

The seventh month of year is ................

The eighth month of year is ................

The ninth month of year is ................

The tenth month of year is ................

The eleventh month of year is ................

The twelfth month of year is ................

.